AF429178

After suffering an attack against his exemplary and impeccable professional career, Engineer Israel Laisequilla offers us in this work a detailed and clearly written guide that allows us to delve into the world of industry in the unique manner that only the so-called "most controversial engineer" can achieve.

INSTRUCTIONS

The information presented here takes into account the access and/or availability of current information and technologies. In case you require more information, formats, and/or examples, your inquiry may increase its reliability thanks to the criteria developed with the book. Any additional information, formulas, and/or videos may be requested using your preferred artificial assistant.

the all about
Lean Six Sigma

ENGR'S WORKSHOP

I. LAISEQUILLA

the all about Lean Six Sigma

Revision 1

Author's introduction added, July 2024.

ii

Dedicated to all the curious minds who seek to understand the world through data.

the all about Lean Six Sigma

CONTENT

Acknowledgements   vi

1   Introduction to Lean Six Sigma   1

2   History and evolution of Lean Six Sigma   4

3   Principles of Lean Six Sigma   7

4   Structure and components of Lean Six Sigma   11

5   Roles and responsibilities of a Lean Six Sigma team   16

6   Project selection in Lean Six Sigma   20

7   Problem definition and measurement   24

8   Data analysis and statistical tools   27

9   Process improvement using Lean Six Sigma   31

10   Methods of control and quality assurance in Lean Six Sigma   34

11   Implementation of Lean Six Sigma in an organization   38

12   Communication and leadership in Lean Six Sigma projects   45

13   Identification and waste reduction in processes   49

14   Identification and elimination of process bottlenecks   53

15   Design of experiments and analysis of variance   60

16   Development of innovative solutions using Lean Six Sigma   64

17   Identification and analysis of risks in processes   67

18   Value chain analysis and delivery time reduction   71

19   Implementation of quality management systems   75

20   Integration of Lean Six Sigma with other process improvement methodologies   80

21   Problem-solving complex problems using Lean Six Sigma   84

22   Lean Six Sigma project management   89

23   Development and management of Lean Six Sigma teams   93

24   Training and certification of Lean Six Sigma professionals   96

25   Implementation of Lean Six Sigma in different industrial sectors: 101
applications and success cases

ACKNOWLEDGEMENTS

Before delving into the world of Lean Six Sigma, I would like to take a moment to express my gratitude to the people who made this book possible.

First and foremost, I want to thank my editor, who provided me with support and guidance throughout the entire writing process. Their expertise and dedication were invaluable in bringing this project to fruition.

I also want to thank my family, who have supported me unconditionally in all of my endeavors, including this one. Their words of encouragement and infinite patience have allowed me to dedicate the time and energy necessary to complete this work.

Last but not least, I want to thank all the people who seek to understand the world around us through data and statistical analysis. Their passion for discovering patterns and truths in information has been a constant source of inspiration for me. Thanks to their curiosity, this book has come to life.

Thank you once again to everyone for your support and trust. I hope you enjoy reading this book as much as I enjoyed writing it.

engr's Workshop

INTRODUCTION TO LEAN SIX SIGMA

Lean Six Sigma is a continuous improvement methodology that combines the principles of Lean Manufacturing and Six Sigma to eliminate waste and reduce variability in processes. This methodology aims to improve the quality of products and services, reduce costs, and increase customer satisfaction.

Lean Manufacturing focuses on eliminating waste, while Six Sigma focuses on reducing variability. By combining these two methodologies, Lean Six Sigma seeks to improve process efficiency and effectiveness holistically.

The Lean Six Sigma methodology consists of five phases: Define, Measure, Analyze, Improve, and Control (DMAIC). Each of these phases has a specific purpose and is used to carry out continuous improvement in processes.

Phase 1: Define

The Define phase aims to define the problem and establish improvement objectives. During this phase, the processes to be improved are identified, and improvement objectives are established in terms of quality, cost, and time.

A improvement team is also created to carry out the project. This team should be composed of individuals from different areas of the organization to have a broad and diverse perspective.

Phase 2: Measure

The Measure phase aims to collect data about the current process to have a baseline for comparison with the improved process. During this phase, critical variables that affect process quality are identified, and performance indicators to measure improvement are established.

Tools and techniques are also established to collect and analyze data. Some of these tools include flowcharts, Pareto charts, histograms, and control charts.

Phase 3: Analyze

The Analyze phase aims to identify the root causes of the problem and find improvement opportunities. During this phase, the data collected in the Measure phase is analyzed to identify patterns and trends.

Analysis tools and techniques such as root cause analysis, correlation analysis, and variance analysis are used to identify the causes of the problem and improvement opportunities.

Phase 4: Improve

The Improve phase aims to implement solutions to address the root causes identified in the Analyze phase. During this phase, potential solutions are developed, and the best solution is selected for implementation.

The design of experiments tool is also used to test the solution and ensure its effectiveness before implementing it in the process.

Phase 5: Control

The Control phase aims to sustain the improvement and prevent the recurrence of the problem. During this phase, control measures are established to ensure that the implemented solution is maintained over time.

Follow-up and monitoring plans are also established to measure the effectiveness of the solution and take corrective actions if necessary. Feedback systems and periodic evaluations can be established to ensure that the process continues to improve over time.

In addition to the five DMAIC phases, the Lean Six Sigma methodology also includes other tools and techniques used to improve quality and reduce costs. Some of these tools include:

Value stream maps: This tool is used to visualize the entire process and identify areas of waste. It allows for the identification of improvement opportunities and the development of a plan to eliminate waste and improve process flow.

5S: This technique is used to organize the workplace and improve efficiency. It involves sorting, setting in order, cleaning, standardizing, and sustaining the workplace.

Poka-yoke: This technique is used to prevent errors and defects in the process. It involves designing the process in a way that makes it impossible to make errors or defects.

Kanban: This technique is used to control the flow of work and improve efficiency. It involves using cards or visual signals to indicate when more products should be produced or when inventory should be replenished.

The Lean Six Sigma methodology can be applied in any type of organization, from manufacturing companies to service organizations. It can also be applied to any process, from product manufacturing to customer service.

The success of Lean Six Sigma implementation depends on the organization's culture. It is important for the organization to have a culture of continuous improvement and be committed to implementing Lean Six Sigma. It is also important for organizational leaders to support and actively participate in the continuous improvement process.

In summary, Lean Six Sigma is a continuous improvement methodology that combines the principles of Lean Manufacturing and Six Sigma to eliminate waste and reduce variability in processes. This methodology aims to improve the quality of products and services, reduce costs, and increase customer satisfaction. The methodology consists of five phases: Define, Measure, Analyze, Improve, and Control (DMAIC), and utilizes tools and techniques such as value stream maps, 5S, Poka-yoke, and Kanban to improve process efficiency and effectiveness. The success of Lean Six Sigma implementation depends on the organization's culture and the commitment of leaders.

HISTORY AND EVOLUTION OF LEAN SIX SIGMA

In recent years, Lean Six Sigma has become a popular methodology for continuous improvement that has been used in various organizations and businesses. This methodology combines the principles of Lean and Six Sigma to eliminate waste, reduce defects, and improve the efficiency and effectiveness of processes. In this chapter, we will discuss the history and evolution of Lean Six Sigma, from its origins to its current adoption in the business world.

Origins of Lean Six Sigma

The Lean philosophy originated in Japan in the 1950s, in the automotive industry, specifically at Toyota. The primary goal of Lean is to maximize customer value while minimizing waste. The main focus of Lean is the elimination of waste in processes, such as overproduction, waiting time, transportation, over-processing, inventory, motion, and defects.

On the other hand, Six Sigma is a methodology that originated at Motorola in the 1980s. The primary goal of Six Sigma is to reduce variability and defects in processes. The methodology uses statistical tools to identify and analyze problems and improve the quality of products and services.

The combination of Lean and Six Sigma occurred in the 1990s when companies began to seek ways to improve their efficiency and effectiveness in processes. The Lean Six Sigma methodology became a popular tool for companies to improve their quality, reduce costs, and enhance customer satisfaction.

Development of Lean Six Sigma

The Lean Six Sigma methodology has evolved over time to adapt to the changing needs of businesses. Below are some key stages in the evolution of Lean Six Sigma:

Emergence of Lean Six Sigma methodology

The Lean Six Sigma methodology began to be used in the 1990s in companies such as General Electric and Motorola. At this time, the methodology focused on process improvement and cost reduction.

The Lean Six Sigma methodology is based on five phases, known as DMAIC (Define, Measure, Analyze, Improve, and Control). Each phase has a set of tools and techniques used to identify and solve problems in processes.

Expansion of Lean Six Sigma to other industries

In the 2000s, the Lean Six Sigma methodology expanded to other industries, such as healthcare and government. Companies began to use the methodology to improve the quality of services and reduce costs.

The Lean Six Sigma methodology was also adapted to address specific issues in each industry. For example, in the healthcare industry, Lean Six Sigma is used to improve process efficiency and reduce medical errors.

Integration of Lean Six Sigma with other methodologies

In the last decade, the Lean Six Sigma methodology has been integrated with other methodologies, such as design thinking and agile project management. The integration of these methodologies has led to the creation of new tools and techniques used in continuous improvement.

The Lean Six Sigma methodology has also been adapted to address specific issues within each organization. Customized versions of Lean Six Sigma have been developed for the service industry, the public sector, and the nonprofit sector. A version for small and medium-sized enterprises has also been created.

Furthermore, the Lean Six Sigma methodology has evolved to include the concept of Lean Startup. The Lean Startup methodology focuses on creating products and services that meet customer needs. It is used to develop minimum viable products (MVPs) and experiment with the market to obtain early feedback.

Benefits of Lean Six Sigma

The Lean Six Sigma methodology is used to improve quality, reduce costs, and increase customer satisfaction. The benefits of Lean Six Sigma include:

Quality improvement: Lean Six Sigma helps improve the quality of products and services by reducing defects and variability in processes.

Cost reduction: Lean Six Sigma helps reduce costs by eliminating waste in processes and improving efficiency.

Increased customer satisfaction: Lean Six Sigma helps improve customer satisfaction by enhancing the quality of products and services and reducing wait times.

Efficiency improvement: Lean Six Sigma helps improve efficiency by eliminating waste and enhancing processes.

Increased profitability: Lean Six Sigma helps increase profitability by reducing costs and improving the quality of products and services.

Conclusion

The Lean Six Sigma methodology has evolved over time to adapt to the changing needs of businesses and organizations. The combination of Lean and Six Sigma principles has proven effective in improving quality, reducing costs, and increasing customer satisfaction. The methodology has been adapted to address specific issues in each industry and organization. The Lean Six Sigma methodology will continue to evolve and adapt to meet the future needs of businesses and organizations worldwide.

PRINCIPLES OF LEAN SIX SIGMA

Lean Six Sigma is a management approach that combines Lean and Six Sigma methodologies to improve quality, reduce costs, and increase efficiency in an organization. This approach focuses on waste elimination, continuous improvement, and reduction of variability in production processes. This chapter will focus on the basic principles of Lean Six Sigma and how they can be applied in an organization to achieve better results.

Introduction to Lean Six Sigma

Lean Six Sigma is a management approach that has been used in many organizations to improve quality, reduce costs, and increase efficiency. This methodology combines the tools and techniques of Lean and Six Sigma to create a continuous improvement approach within the organization. The principles of Lean Six Sigma are based on waste elimination, continuous improvement, and reduction of variability in production processes.

Principles of Lean Six Sigma

The principles of Lean Six Sigma are based on five key principles: customer focus, continuous improvement, waste elimination, variability reduction, and performance improvement. Each of these principles is described below.

Customer Focus

The first principle of Lean Six Sigma is customer focus. The idea is that an organization should focus on what the customer wants and needs, rather than focusing on what the organization wants to produce. To achieve this, an organization must understand its customers' needs and ensure that its products and services meet those needs.

Continuous Improvement

The second principle of Lean Six Sigma is continuous improvement. The idea is that an organization should always strive to improve its processes and products to better meet its customers' needs. This is achieved through the identification and elimination of waste, the reduction of variability, and the increase in efficiency.

Waste Elimination

The third principle of Lean Six Sigma is waste elimination. The idea is that an organization should eliminate any activity or process that does not add value to the product or service being produced. This is achieved through the identification and elimination of redundant processes, unnecessary activities, and anything else that does not contribute to the final value of the product or service.

Variability Reduction

The fourth principle of Lean Six Sigma is variability reduction. The idea is that an organization should minimize variability in its processes to achieve greater efficiency and quality. This is achieved through the identification and elimination of any source of variability in the production process.

Performance Improvement

The fifth principle of Lean Six Sigma is performance improvement. The idea is that an organization should constantly measure and improve its performance to achieve greater efficiency and quality. This is achieved through the monitoring and measurement of key performance indicators and the implementation of improvements in production processes.

How to Apply the Principles of Lean Six Sigma

Applying the principles of Lean Six Sigma in an organization requires a structured and systematic approach. The necessary steps to apply these principles in an organization are described below.

Identify Key Processes

The first step in applying the principles of Lean Six Sigma is to identify the key processes in the organization that need improvement. This can be achieved by conducting an analysis of production processes, identifying those with higher costs, longer cycle times, or quality issues.

Analyze Processes

Once the key processes have been identified, a detailed analysis of these processes is necessary to identify areas for improvement. This can be achieved by creating process flow diagrams, identifying critical points, and pinpointing quality issues.

Identify Waste

Once the problems and areas for improvement in the processes have been identified, it is necessary to identify waste within them. This can be achieved by identifying activities that do not add value to the final product or service and eliminating them.

Reduce Variability

Once waste has been identified, it is necessary to reduce variability in the processes to achieve greater efficiency and quality. This can be achieved by identifying and eliminating sources of variability in the production process.

Measure and Improve Performance

Finally, it is necessary to constantly measure and improve the performance of the processes to achieve greater efficiency and quality. This can be achieved through the monitoring and measurement of key performance indicators and the implementation of improvements in production processes.

Benefits of Applying the Principles of Lean Six Sigma

Applying the principles of Lean Six Sigma in an organization can offer several benefits, including:

Improved product or service quality.

Reduced production costs.

Increased efficiency and productivity.

Improved customer satisfaction.

Reduced cycle time of production processes.

Increased profitability of the organization.

Conclusions

Lean Six Sigma is a management approach that combines Lean and Six Sigma methodologies to improve quality, reduce costs, and increase efficiency in an organization. The principles of Lean Six Sigma are based on waste elimination, continuous improvement, and reduction of variability in production processes. Applying these principles requires a structured and systematic approach, including the identification of key processes, analysis, waste identification, variability reduction, and performance measurement and improvement. Applying the principles of Lean Six Sigma can provide several benefits to an organization, including improved quality, cost reduction, increased efficiency and productivity, improved customer satisfaction, and profitability.

STRUCTURE AND COMPONENTS OF LEAN SIX SIGMA

Lean Six Sigma is a methodology that combines two approaches to improve business processes and reduce errors: Lean Manufacturing and Six Sigma. The methodology focuses on waste elimination and variation reduction in processes. In this chapter, we will analyze the structure and components of Lean Six Sigma, which include the Lean philosophy, the Six Sigma methodology, and the tools used in the implementation of this methodology.

Lean Philosophy

The Lean philosophy originated in Japan and focused on eliminating waste in business processes. Waste refers to anything that does not add value to the product or service, such as waiting time, excess inventory, overproduction, unnecessary transportation, overprocessing, and defects. By eliminating these wastes, efficiency is improved, and costs are reduced.

The Lean philosophy is based on five principles:

Value: The customer determines the value of a product or service.

Value Stream: Processes should be designed to create a continuous value stream from raw material to the customer.

Flow: Processes should be designed to minimize waste and maximize the value flow.

Pull: Only produce what is needed, when it is needed, and in the necessary quantity.

Perfection: The ultimate goal is perfection, meaning the complete elimination of waste.

The Lean philosophy focuses on continuous improvement and organizational learning. Employees are encouraged to identify and eliminate waste in their processes and constantly seek ways to improve. This philosophy applies not only to manufacturing but also to services and other industries.

Six Sigma Methodology

The Six Sigma methodology focuses on reducing variation in business processes. Variation refers to any deviation from the standard process that can lead to errors or defects in the product or service. The methodology uses a set of statistical tools and techniques to identify and reduce variation in processes.

The Six Sigma methodology is based on five phases:

Define: In this phase, the problem is defined, and a project team is established.

Measure: Data is collected to understand the process and quantify the variation.

Analyze: Data is analyzed to identify the root causes of variation.

Improve: Solutions are developed to reduce variation, and the changes are implemented.

Control: Controls are established to ensure that the changes are sustained, and results are measured to ensure continuous improvement.

The Six Sigma methodology employs a data-driven and fact-based approach to decision-making and focuses on customer satisfaction and continuous improvement. The goal is to reduce variation to a six sigma level, which means the process produces only 3.4 defects per million opportunities.

Lean Six Sigma Tools

The Lean Six Sigma methodology uses a variety of tools and techniques to identify and eliminate waste and reduce variation in processes. Below are some of the most common tools used in Lean Six Sigma:

Value Stream Map: A value stream map is a visual tool that shows the flow of materials and information in a process. It is used to identify waste in a process and design a more efficient value stream.

Pareto Analysis: Pareto analysis is a technique used to identify the most common problems in a process. It is based on the principle that 80% of the problems come from 20% of the causes.

Ishikawa Diagram: Also known as a fishbone diagram, it is a tool used to identify possible root causes of a problem. It is used to identify different categories of causes that may contribute to a problem.

Histogram: A histogram is a graph that shows the distribution of data. It is used to identify variation in a process and determine if the data follows a normal distribution.

Control Chart: A control chart is a tool used to monitor a process and detect any unwanted variation. It is used to identify when a process is out of control and take corrective measures.

Process Capability Analysis: Process capability analysis is used to measure a process's ability to meet customer specifications. It is used to identify if a process is producing products or services that meet customer requirements.

DMAIC: DMAIC is a tool used in the Six Sigma methodology to identify and reduce variation in a process. The five phases of DMAIC are described earlier in this chapter.

Roles and Responsibilities in Lean Six Sigma

To successfully implement Lean Six Sigma in an organization, it is important to assign clear roles and responsibilities to employees and company leaders. Below are some common roles in a Lean Six Sigma team:

Project Leader: The project leader is responsible for leading the project team and ensuring project objectives are met.

Project Sponsor: The project sponsor is an executive leader in the organization who provides support and resources to the project team.

Lean Six Sigma Champion: The Lean Six Sigma champion is a leader who supports the implementation of Lean Six Sigma in the organization and acts as an advocate for the methodology.

Black Belt: The Black Belt is a Lean Six Sigma expert who leads complex projects and works with other team members to implement solutions.

Green Belt: The Green Belt is a project team member who works on less complex projects and provides support to the Black Belt.

Yellow Belt: The Yellow Belt is a project team member with basic knowledge of Lean Six Sigma and can assist with specific tasks in a project.

Project Team: The project team consists of individuals from different areas of the organization working together to achieve project objectives. Each team member has a specific role and assigned responsibilities.

It is important to note that implementing Lean Six Sigma requires a cultural change in the organization. All employees must be willing to learn and adapt to new processes and techniques. Organizational leaders must provide a supportive and motivating environment to ensure successful implementation.

Benefits of Lean Six Sigma

Implementing Lean Six Sigma can provide several benefits to an organization. Some common benefits of Lean Six Sigma include:

Cost Reduction: Waste elimination and variation reduction in processes can help reduce production costs and improve organizational efficiency.

Quality Improvement: By reducing variation in processes, the organization can produce higher quality products and services that better meet customer needs.

Increased Customer Satisfaction: Improved quality and efficiency can increase customer satisfaction and enhance the organization's reputation.

Increased Productivity: Waste elimination and variation reduction in processes can increase organizational productivity.

Enhanced Innovation Capability: By improving efficiency and quality, the organization can allocate more resources to innovation and the development of new products and services.

Improved Organizational Culture: Implementing Lean Six Sigma can improve the organizational culture by fostering collaboration, innovation, and continuous learning.

Conclusion

In summary, Lean Six Sigma is a powerful methodology that can help organizations improve efficiency, reduce costs, and enhance the quality of their products and services. The methodology is based on waste elimination and variation reduction in processes and utilizes a variety of tools and techniques to achieve these objectives.

To successfully implement Lean Six Sigma in an organization, it is important to assign clear roles and responsibilities to employees and company leaders. Additionally, implementing Lean Six Sigma requires a cultural change in the organization, and all employees must be willing to learn and adapt to new processes and techniques.

The benefits of Lean Six Sigma include cost reduction, quality improvement, increased customer satisfaction, increased productivity, enhanced innovation capability, and improved organizational culture.

Ultimately, the implementation of Lean Six Sigma can help organizations achieve a competitive advantage in the market and improve their position in the industry.

ROLES AND RESPONSIBILITIES OF A LEAN SIX SIGMA TEAM

Lean Six Sigma is a continuous improvement methodology that focuses on waste elimination and reduction of variability in processes. This methodology relies on collaboration among teams led by project leaders and supported by experts in data analysis and statistical tools. This chapter describes the roles and responsibilities of members in a Lean Six Sigma team, including project leaders and team members. It also discusses the role of data analysis experts and statistical tools.

Roles and Responsibilities of a Project Leader

The project leader is responsible for leading the Lean Six Sigma team to achieve project objectives. Their main responsibilities include:

Defining project scope: The project leader must clearly and concisely define the project scope, ensuring that all team members have a common understanding of what is expected to be achieved. The project scope should include objectives, goals, and timelines.

Identifying the team: The project leader should select the appropriate team members for the project. Team members should have complementary skills and knowledge that enable them to work effectively together.

Facilitating communication: The project leader should facilitate communication among team members and other stakeholders involved in the project. Communication should be clear and effective, and the project leader must ensure that all team members are aligned with the project objectives.

Establishing a project plan: The project leader should develop a project plan that includes tasks, deadlines, and resources required to achieve project objectives. The project plan should be realistic and consider available resources and limitations.

Leading the team: The project leader must effectively lead the team, motivating team members and ensuring that project deadlines and objectives are met. The project leader should also ensure that team members are working together effectively.

Measuring and monitoring project progress: The project leader should measure and monitor project progress against established timelines and objectives. The project leader must identify and address any issues that may hinder project success.

Presenting project results: The project leader should present project results to stakeholders. The presentation should be clear and concise, highlighting achievements and lessons learned during the project.

Roles and Responsibilities of Team Members Team members are responsible for carrying out tasks assigned by the project leader and working together to achieve project objectives. Their main responsibilities include:

Contributing to the project plan: Team members should contribute to the project plan established by the project leader. Team members should ensure that their tasks are aligned with project objectives and inform the project leader of any issues or challenges that may affect project progress.

Performing assigned tasks: Team members should complete assigned tasks within established deadlines and in an effective manner. Team members should ensure that their tasks are complete and well-documented.

Working as a team: Team members should work together effectively and collaboratively. Team members should communicate effectively and promptly resolve any issues or conflicts.

Identifying and addressing issues: Team members should identify and address any issues or challenges that may hinder project progress. Team members should work together to find solutions and take corrective actions.

Providing data and analysis: Team members should provide relevant data and analysis to support the project. Team members should ensure that the data and analysis are accurate and well-documented.

Participating in the presentation of results: Team members should participate in the presentation of project results. Team members should be prepared to present their work and respond to questions related to the project.

Roles and Responsibilities of data analysis experts and statistical tools. Data analysis experts and statistical tools are responsible for providing guidance and technical support to the project leader and team members. Their main responsibilities include:

Identifying and applying statistical tools: Data analysis experts should identify and apply relevant statistical tools for the project. They should ensure that statistical tools are applied effectively and appropriately.

Providing data analysis: Data analysis experts should provide relevant data analysis to support the project. They should ensure that the analysis is accurate and well-documented.

Providing technical advice: Data analysis experts should provide technical advice to the project leader and team members. They should ensure that the project leader and team members understand the application of statistical tools and data analysis.

Participating in the presentation of results: Data analysis experts should participate in the presentation of project results. They should be prepared to present their work and respond to questions related to the project.

Conclusion

In summary, the roles and responsibilities of members in a Lean Six Sigma team are critical to project success. The project leader is responsible for leading the team and ensuring project objectives are met. Team members must work together effectively and complete assigned tasks. Data analysis experts and statistical tools are responsible for providing guidance and technical support to the project leader and team members.

PROJECT SELECTION IN LEAN SIX SIGMA

Project selection is a crucial stage in the implementation process of Lean Six Sigma (LSS). During this stage, critical problems that affect quality, costs, and delivery times within an organization are identified. Projects that have the highest impact are prioritized, and improvement objectives are defined. The success of LSS implementation greatly depends on appropriate project selection.

This chapter describes the process of project selection in LSS, presents the tools and techniques used for project identification and prioritization, discusses selection criteria, and provides practical examples.

Process of Project Selection in LSS

The process of project selection in LSS consists of several stages, as described below:

Identification of critical problems: The first stage in the project selection process is to identify critical problems that affect quality, costs, and delivery times within an organization. These problems can be identified using various tools and techniques such as data analysis, process analysis, root cause analysis, among others.

Prioritization of projects: Once the critical problems have been identified, it is necessary to prioritize the projects that have the highest impact on the organization. Various tools and techniques can be used for this purpose, including prioritization matrices, cost-benefit analysis, risk analysis, among others.

Definition of improvement objectives: After prioritizing the projects, it is essential to define improvement objectives for each selected project. These objectives should be Specific, Measurable, Achievable, Relevant, and Time-bound (SMART). Improvement objectives must align with the organization's strategy and contribute to the enhancement of key performance indicators (KPIs).

Selection of a working team: Once the improvement objectives are defined, a working team needs to be selected for each project. This team should consist of individuals with specific skills and knowledge to address the identified problem and achieve the defined improvement objectives.

Tools and Techniques for Project Identification and Prioritization

There are various tools and techniques that can be used for project identification and prioritization in LSS. Some of the most common ones are described below:

Data analysis: Data analysis is a key tool for identifying critical problems within an organization. Various statistical techniques such as variance analysis, regression analysis, trend analysis, among others, can be used to identify patterns and trends in the data that indicate the presence of critical problems.

Process analysis: Process analysis is used to identify critical problems in production or service processes. Various tools such as flowcharts, process maps, value stream analysis, among others, can be employed to identify bottlenecks, inefficiencies, and improvement opportunities within processes.

Root cause analysis: Root cause analysis is used to identify the underlying causes of critical problems. Different tools such as fishbone diagrams, 5 Whys analysis, among others, can be utilized to identify the root causes of problems.

Prioritization matrix: A prioritization matrix is a tool that enables the comparison and prioritization of different projects based on their impact on the organization and their feasibility of implementation. Various criteria such as impact on quality, impact on costs, impact on delivery times, implementation complexity, among others, can be used to compare and prioritize projects.

Cost-benefit analysis: Cost-benefit analysis is used to evaluate the profitability of projects. Implementation costs are compared with expected benefits, and it is determined whether the project is profitable. This tool is particularly useful for evaluating improvement projects involving significant investments.

Risk analysis: Risk analysis is used to assess the risks associated with improvement projects. Potential risks are identified, their probability of occurrence and impact on the project are evaluated, and contingency plans are defined to mitigate the risks.

Criteria for Project Selection in LSS

To select the appropriate LSS projects, several criteria need to be considered, including the following:

Impact on the organization: The selected projects should have a significant impact on the organization. Projects that have a higher impact on quality, costs, and delivery times should be prioritized.

Feasibility of implementation: The selected projects should be feasible to implement. Factors such as implementation complexity, resource availability, and time required for project implementation should be taken into account.

Alignment with the organization's strategy: The selected projects should align with the organization's strategy. Projects that contribute to achieving the organization's strategic objectives should be prioritized.

Profitability: The selected projects should be profitable. Implementation costs and expected benefits should be evaluated, and projects with appropriate profitability should be selected.

Practical Examples

The following are some practical examples of LSS project selection:

Reducing delivery times: In a manufacturing company, product delivery times are long and unpredictable. A process analysis is conducted, revealing various inefficiencies in the production process. A project is selected to reduce delivery times through the implementation of process improvements. SMART objectives are defined for the project, a working team is selected, and the identified process improvements are implemented.

Improving product quality: In a technology services company, a high error rate has been identified in the software development process. A process analysis is conducted, uncovering various inefficiencies in the software development process. A project is selected to improve product quality through the implementation of process improvements. SMART objectives are defined for the project, a working team is selected, and the identified process improvements are implemented.

Cost reduction: In a financial services company, operational costs are excessively high. A process analysis is conducted, identifying various inefficiencies in operational processes. A project is selected to reduce costs through the implementation of process improvements. SMART objectives are defined for the project, a working team is selected, and the identified process improvements are implemented.

Conclusion

Appropriate project selection in LSS is essential for the success of any continuous improvement initiative within an organization. To select the right projects, a rigorous analysis of processes needs to be performed, inefficiencies and improvement opportunities must be identified, and appropriate selection criteria should be evaluated. Furthermore, it is important to define SMART objectives, select a skilled and committed working team, and ensure the availability of necessary resources for project implementation. By following these best practices, organizations can improve their performance, increase profitability, and strengthen their market position.

PROBLEM DEFINITION AND MEASUREMENT

Lean Six Sigma is a management approach that focuses on waste elimination and variation reduction in a company's processes. To achieve this, specific tools and techniques are used to identify and solve problems within the organization. This chapter will address the definition and measurement of problems in Lean Six Sigma, which is essential for the success of any continuous improvement initiative.

Definition of Problems in Lean Six Sigma

Before delving into the definition of problems in Lean Six Sigma, it is important to understand that the word "problem" can have different meanings depending on the corporate culture and context. In Lean Six Sigma, a problem is defined as any situation that prevents a company from achieving its objectives or causes dissatisfaction among customers. In other words, a problem is any gap between what is expected from a process and what is actually obtained.

Problems in Lean Six Sigma are classified into three main categories: chronic problems, acute problems, and improvement opportunities. Chronic problems are those that occur consistently and have a significant impact on the company's performance. Acute problems, on the other hand, are those that arise unexpectedly and require immediate action to prevent further damage. Lastly, improvement opportunities are situations where there is potential to enhance a process, but there is no real problem to be solved.

Measurement of Problems in Lean Six Sigma

Once a problem has been defined in Lean Six Sigma, it is important to measure it in order to understand its impact on the company's performance and establish a starting point for improvement. The measurement of problems is carried out through different techniques and tools, which vary depending on the type of problem and the industry in which the company operates.

In general, the measurement of problems in Lean Six Sigma is carried out in three phases: measurement of the current problem, root cause analysis, and measurement of solution impact. The following are details for each of these phases.

Measurement of the Current Problem

Measuring the current problem is the first phase in the measurement of problems in Lean Six Sigma. In this phase, data is collected to understand the impact of the problem on the company's performance. Data can be gathered from various sources such as the quality management system, the company's information system, customer and employee surveys, among others.

A commonly used tool in this phase is the Pareto chart, which helps identify the problems that have a significant impact on the company's performance. The Pareto chart is constructed by ordering the problems from highest to lowest impact and plotting the frequency of each problem on one axis and the cumulative impact on the other axis.

Root Cause Analysis

Once the current problem has been measured, the next step is to conduct a root cause analysis. In this phase, the underlying causes of the problem are sought in order to address them effectively. Root cause analysis can be performed using different tools, such as the Five Whys analysis, which involves asking successive questions about the problem until reaching its root cause. Advanced tools like the fishbone diagram or Ishikawa diagram can also be used to identify different causes of the problem and group them into categories.

Once the root causes of the problem have been identified, a cost-benefit analysis approach can be used to determine the best solution to implement. This approach compares the cost of implementing the solution with the benefits that will be obtained, ensuring that the chosen solution is cost-effective.

Measurement of Solution Impact

The final phase in the measurement of problems in Lean Six Sigma is the measurement of solution impact. Once the solution has been implemented, the results need to be measured to ensure that the problem has been effectively resolved and improvements in the company's performance have been achieved.

The measurement of solution impact is carried out using the same indicators that were used in the measurement of the current problem. If sufficient data has been collected during the measurement of the current problem phase, the data before and after implementing the solution can be compared to determine the impact of the improvement.

Conclusions

The definition and measurement of problems in Lean Six Sigma are fundamental for the success of any continuous improvement initiative in a company. Defining problems enables the company to identify gaps between expected and actual process outcomes, while measuring problems helps establish a starting point for improvement and measure the impact of the solution.

The measurement of problems in Lean Six Sigma is carried out in three phases: measurement of the current problem, root cause analysis, and measurement of solution impact. Each of these phases utilizes different tools and techniques to achieve its objective, enabling a structured and efficient approach to problem-solving.

In summary, the definition and measurement of problems in Lean Six Sigma are powerful tools for any company seeking to improve its performance and reduce costs. By effectively utilizing these tools, a company can proactively identify and solve problems, leading to increased efficiency and competitiveness in the market.

DATA ANALYSIS AND STATISTICAL TOOLS

In recent years, the use of tools and methodologies for data analysis has become a common practice in companies seeking to improve their efficiency and effectiveness. One of these methodologies is Lean Six Sigma, which focuses on improving business processes through waste elimination and variability reduction in production.

This chapter will address data analysis and statistical tools used in Lean Six Sigma to enhance the quality of business processes. It will explain in detail how these tools can be used to identify issues in a process and how they can be applied to reduce variability and improve production efficiency.

Data Analysis in Lean Six Sigma

Data analysis is a fundamental part of Lean Six Sigma as it allows for the identification of issues in a process and finding solutions to enhance quality and efficiency. In this process, statistical tools and techniques are employed to analyze data and identify patterns that enable informed decision-making on process improvement.

Data collection is the first stage of data analysis in Lean Six Sigma. Data can be collected in various ways, including direct process observation, record review, and conducting surveys. Once the data has been collected, it can be analyzed using different statistical techniques.

A common technique used in Lean Six Sigma for data analysis is Pareto analysis.

This technique is used to identify the most common issues in a process and determine their root causes. Pareto analysis is based on the 80/20 principle, which states that 80% of the problems are caused by 20% of the causes. By identifying and addressing these root causes, most of the process problems can be resolved.

Another technique used in data analysis in Lean Six Sigma is correlation analysis. This technique is employed to identify the relationship between two variables in a process. For example, if there is a suspicion that production speed is related to the number of errors in the process, correlation analysis can be used to determine if there is a relationship between these two variables. If a positive correlation is found, meaning that as production speed increases, errors also increase, measures can be taken to reduce production speed and improve process quality.

Regression analysis is another statistical technique used in Lean Six Sigma for data analysis. This technique is employed to predict the value of a variable based on other variables. For instance, if one wants to predict the number of defective products produced in a week, regression analysis can be used to determine if there is a relationship between the number of working hours and the number of defective products produced. If a significant relationship is found, measures can be taken to reduce working hours and improve process quality.

Other techniques used in data analysis in Lean Six Sigma include analysis of variance, time series analysis, and capability analysis. These techniques allow for data analysis in different ways to identify patterns and trends in the process, which can aid in making informed decisions on how to improve quality and efficiency.

Statistical Tools in Lean Six Sigma

In addition to the aforementioned statistical techniques, Lean Six Sigma employs a range of specific tools to analyze data and improve business processes. Here are some of these tools:

Ishikawa diagram

The Ishikawa diagram, also known as the fishbone diagram or cause-and-effect diagram, is a tool used in Lean Six Sigma to identify the root causes of a problem in a process. The diagram is used to visualize the different causes that may contribute to a problem and determine the most significant ones. Causes are grouped into different categories such as labor, machinery, materials, methods, environment, and measurement. The Ishikawa diagram allows for structured problem analysis and finding solutions to resolve it.

Flowchart

The flowchart is another tool used in Lean Six Sigma to analyze a process and determine areas where efficiency and quality can be improved. The diagram is used to visualize the flow of the process, from input of materials to the output of the final product. Each stage of the process is represented with a symbol and connected with arrows to indicate the flow. The flowchart enables the identification of areas where waste can be reduced and efficiency can be improved in the process.

Control chart

The control chart is a tool used in Lean Six Sigma to monitor a process and detect any changes that may indicate variability in the process. The chart is used to visualize process data over time and determine if the data is within established control limits. If the data deviates from the control limits, measures can be taken to correct the process and improve quality and efficiency.

Correlation matrix

The correlation matrix is a tool used in Lean Six Sigma to visualize the relationship between multiple variables in a process. The matrix is used to visualize the correlation between different variables and determine the most important ones in the process. The correlation matrix allows for the identification of variables that need to be monitored and controlled to improve quality and efficiency in the process.

Capability analysis

Capability analysis is a tool used in Lean Six Sigma to determine the ability of a process to meet customer specifications. The analysis is used to compare process variability with customer specifications and determine if the process is producing products within specifications. If the process fails to meet customer specifications, measures can be taken to reduce variability and improve process quality.

Conclusions

In conclusion, data analysis and statistical tools are crucial in Lean Six Sigma to enhance quality and efficiency in business processes. The different statistical techniques and tools enable the identification of root causes, reduction of process variability, optimization of efficiency, and improvement of customer satisfaction.

It is important to note that successfully implementing Lean Six Sigma requires a highly trained team proficient in the application of these statistical techniques and tools. Furthermore, it is essential for the organization to have a customer-centric focus and a commitment to continuous process improvement.

In summary, data analysis and statistical tools are a key component in Lean Six Sigma and are essential for problem identification and resolution, variability reduction, and efficiency improvement in business processes. When applied correctly, these techniques can assist organizations in enhancing their performance and remaining competitive in the market.

PROCESS IMPROVEMENT USING LEAN SIX SIGMA

Process improvement is a critical aspect of business management and competitiveness in today's market. Inefficient and ineffective business processes can cause delays, errors, customer dissatisfaction, and loss of business opportunities. Therefore, it is important for companies to implement strategies and methodologies to continuously improve their processes and increase efficiency and effectiveness.

In this chapter, we will discuss the Lean Six Sigma (LSS) methodology as an effective tool for process improvement. We will explain what LSS entails, how it is applied in process improvement, and provide some examples of successful LSS implementations in different companies and organizations.

What is Lean Six Sigma?

Lean Six Sigma (LSS) is a methodology that combines two approaches: Lean and Six Sigma. Lean focuses on waste elimination and efficiency improvement, while Six Sigma focuses on quality improvement and variability reduction. Together, these approaches offer a comprehensive methodology for improving business processes.

The LSS methodology is based on the PDCA cycle (Plan-Do-Check-Act), which is a systematic approach to continuous process improvement. In this cycle, the process is first planned, executed, the results are checked, and corrective actions are taken to improve the process. The LSS methodology uses specific tools and techniques at each stage of the PDCA cycle to achieve continuous improvement.

Application of Lean Six Sigma in process improvement

The application of the LSS methodology in process improvement involves a series of steps. The typical steps of applying LSS in process improvement are described below:

Problem definition: The first step in applying LSS is to define the problem. It is important to have a clear understanding of the problem and how it affects the business process. In this step, the problem is identified, and improvement objectives are established.

Measurement: In this stage, data is collected to understand the current process and determine its performance. Tools such as value stream analysis, time measurement, and customer satisfaction evaluation are used to identify the process weaknesses.

Analysis: In this stage, the collected data is analyzed to identify the root causes of the problem. Tools such as Ishikawa diagrams and Pareto analysis are used to identify the primary causes of the problem.

Improvement: In this stage, solutions are developed to address the root causes of the problem. Tools such as design of experiments and solution selection matrix are used to develop and select the most effective solutions.

Control: In this stage, the developed solutions are implemented, and the process is monitored to ensure it remains in a state of continuous improvement. Tools such as control charts and process auditing are used to ensure that the process is functioning effectively and stays within established limits.

Examples of Lean Six Sigma Application in process improvement

Below are some examples of how the LSS methodology has been successfully implemented in different companies and organizations:

General Electric: GE has been one of the leaders in LSS implementation. They have implemented LSS throughout the organization, from manufacturing to financial services. As a result, they have been able to reduce costs, improve quality, and increase customer satisfaction.

Ford: Ford has used LSS to improve its car manufacturing processes. They have been able to reduce cycle time in vehicle production, reduce defects, and improve process efficiency.

American Express: American Express has used LSS to improve its customer service process. They have been able to reduce response time to customers, improve customer satisfaction, and reduce service costs.

Amazon: Amazon has used LSS to improve its inventory management and product delivery process. They have been able to reduce product delivery time, reduce unsold inventory, and improve warehouse management efficiency.

Conclusions

The Lean Six Sigma methodology is an effective tool for continuous improvement of business processes. The combination of Lean and Six Sigma allows for a comprehensive approach to process improvement, focusing on both efficiency and quality. The application of LSS in business process improvement involves a series of steps that enable continuous and systematic improvement. Additionally, the LSS methodology has been successfully implemented in different companies and organizations, demonstrating its effectiveness in improving business processes. In summary, the LSS methodology is a valuable tool for companies seeking to continuously improve their processes and increase competitiveness in today's market.

METHODS OF CONTROL AND QUALITY ASSURANCE IN LEAN SIX SIGMA

Control and quality assurance are critical aspects in any continuous improvement process, and it is no different in Lean Six Sigma. The methods used to control and assure quality are essential to ensure that the results are consistent, reliable, and aligned with the project's objectives. In this chapter, we will explore the methods used in Lean Six Sigma for control and quality assurance, including statistical process control, capability analysis, measurement validation, change management, and process review.

Statistical Process Control

Statistical process control (SPC) is a method used to monitor and control variability in a process. SPC is a key tool in Lean Six Sigma as it enables early identification of process problems and corrective actions to be taken before major issues occur. SPC is based on data collection and analysis, using control charts to identify any variations that may be outside acceptable limits.

Several types of control charts are used in Lean Six Sigma, including Xbar-R control charts, Xbar-S control charts, and attribute control charts. Control charts are used to monitor key process variables and determine if the process is within acceptable limits of variability. If any variation outside acceptable limits is detected, corrective actions must be taken to address the issue.

Capability Analysis

Capability analysis is a method used to determine if a process is capable of meeting customer specifications. Capability analysis is used to assess the process's ability to produce products or services within specified limits. Capability analysis is based on the use of statistical data to determine process capability, allowing early identification of process problems and corrective actions to improve process capability.

Several capability indices are used in Lean Six Sigma, including process capability index (Cp), short-term process capability index (Cpk), and long-term process capability index (Ppk). These indices are used to evaluate the process's ability to produce products or services within specified limits. If the process does not meet capability requirements, corrective actions must be taken to improve process capability.

Measurement Validation

Measurement validation is a method used to ensure that measurements taken in the process are accurate and reliable. Measurement validation is essential to ensure reliable and accurate process results. Measurement validation is based on comparing measurements taken with a known reference standard.

Several methods are used in Lean Six Sigma to validate measurements, including repeatability and reproducibility (R&R) studies and linearity studies. R&R studies are used to assess the variation in measurements due to process and operator variation. Linearity studies are used to evaluate measurement accuracy within a specific range of values. If issues with measurement accuracy are detected, corrective actions must be taken to address the problem.

Change Management

Change management is a method used to ensure that process changes are made in a controlled and effective manner. Change management is essential to ensure that process changes do not negatively impact product or service quality. Change management is based on identifying necessary changes, evaluating the impact of the change on the process, and implementing the change in a controlled manner.

Several steps must be followed in change management, including change identification, change impact assessment, change approval, and controlled implementation of the change. If issues with the change are detected, corrective actions must be taken to address the problem.

Process Review

Process review is a method used to assess process performance and determine if project objectives are being met. Process review is used to identify areas that require improvement and determine if the correct methods are being used to control and assure process quality.

Process review is based on data collection and analysis, comparing results with project objectives, and identifying areas that require improvement. If areas requiring improvement are identified, corrective actions must be taken to address the problem.

Conclusions

Control and quality assurance are critical aspects in any continuous improvement process, and Lean Six Sigma is no exception. The methods used to control and assure quality are essential to ensure that the results are consistent, reliable, and aligned with the project's objectives. In this chapter, we explored the methods used in Lean Six Sigma for control and quality assurance, including statistical process control, capability analysis, measurement validation, change management, and process review.

Statistical process control is a key tool in Lean Six Sigma as it enables early identification of process problems and corrective actions to be taken before major issues occur. Capability analysis is used to assess the process's ability to produce products or services within specified limits. Measurement validation is essential to ensure that process results are reliable and accurate. Change management is essential to ensure that process changes do not negatively impact product or service quality. Process review is used to identify areas that require improvement and determine if project objectives are being met.

In general, the effective use of these control and quality assurance methods is essential to ensure the success of continuous improvement projects. Early identification of process problems and taking corrective actions to address them can save valuable time and resources. Additionally, continuously evaluating process performance and identifying areas that require improvement can help ensure efficient and effective project objective achievement.

It is important to note that these methods are not only useful in the context of Lean Six Sigma but can also be applied in a wide range of contexts and projects.

By using these methods, organizations can improve the quality of their products and services, enhance customer satisfaction, and improve the efficiency and effectiveness of their processes.

In summary, control and quality assurance are essential aspects in any continuous improvement project, and Lean Six Sigma is no exception. The methods used in Lean Six Sigma for controlling and assuring quality, such as statistical process control, capability analysis, measurement validation, change management, and process review, are valuable tools to ensure efficient and effective achievement of project objectives. By using these methods, organizations can improve the quality of their products and services, enhance customer satisfaction, and improve the efficiency and effectiveness of their processes.

IMPLEMENTATION OF LEAN SIX SIGMA IN AN ORGANIZATION

The implementation of Lean Six Sigma in an organization is a popular strategy for improving process efficiency and quality. This methodology combines the Lean philosophy, which focuses on waste elimination and increasing efficiency, and Six Sigma, which focuses on reducing variability and improving quality. In this chapter, we will discuss the basics of Lean Six Sigma and explain how this methodology can be implemented in an organization.

What is Lean Six Sigma?

Lean Six Sigma is a process improvement methodology that combines the principles of Lean and Six Sigma. The Lean philosophy focuses on waste elimination and efficiency improvement, while Six Sigma focuses on reducing variability and improving quality. The combination of these two approaches creates a powerful methodology that can significantly enhance process efficiency and quality in an organization.

Benefits of implementing Lean Six Sigma

Implementing Lean Six Sigma can provide several significant benefits to an organization. Some of these benefits include:

Cost reduction: Lean Six Sigma helps identify and eliminate waste in processes, which can result in significant cost reduction.

Quality improvement: Six Sigma focuses on reducing variability and improving quality, which can help reduce defects and enhance customer satisfaction.

Increased efficiency: Lean focuses on improving efficiency and eliminating waste, resulting in faster and less costly processes.

Enhanced customer satisfaction: Improved quality and efficiency can lead to increased customer satisfaction, which can be beneficial for brand image and sales.

Increased employee engagement: Implementing Lean Six Sigma can involve employees in process improvement, improving morale and staff commitment.

Steps for implementing Lean Six Sigma

Implementing Lean Six Sigma involves a series of key steps that should be followed to ensure successful implementation. In this chapter, we will explain the basic steps that should be followed to implement this methodology in an organization.

Identify key processes

The first step in implementing Lean Six Sigma is to identify the key processes in the organization that need improvement. These processes should be carefully selected to ensure that improvements in these processes have a significant impact on overall organization efficiency and quality. It is important to involve relevant members of the organization in this process of selecting key processes as they have a deeper understanding of the processes and can provide valuable insights.

Establish a project team

Once the key processes have been identified, a project team should be established to implement Lean Six Sigma. This team should consist of members from different areas of the organization who have a deep understanding of the selected processes. The team leader should have strong leadership skills and experience in implementing Lean Six Sigma.

Conduct process assessment

The next step is to conduct a process assessment to better understand the challenges and areas for improvement. This may involve data collection, conducting interviews, and process analysis. It is important to involve relevant members of the organization in this process assessment to ensure that all issues and challenges are fully understood.

Set improvement goals and metrics

Once the challenges and areas for improvement have been identified, improvement goals and metrics should be established. These goals should be specific, measurable, attainable, relevant, and timely (SMART) and should reflect the challenges identified in the process assessment. Improvement metrics should be used to measure progress towards the goals and provide feedback on the success of the implementation.

Develop an implementation plan

The next step is to develop a detailed implementation plan that includes the specific actions that need to be taken to achieve the established goals. This plan should include an implementation schedule, budget, and change management plan to ensure that all members of the organization are prepared for the change.

Implement and monitor the plan

Once the implementation plan has been developed, it should be implemented and monitored. It is important to involve the entire project team and other relevant members of the organization in this implementation and monitoring process to ensure that the established goals are achieved.

Evaluate the success of the implementation

Once the implementation plan has been implemented, the success of the implementation should be evaluated. This may involve conducting a follow-up assessment to measure progress towards the established goals and identify any areas that need further improvement. It is important to use the improvement metrics established in step 2.4 to evaluate the success of the implementation.

Tips and Suggestions for Successful Implementation of Lean Six Sigma

Successful implementation of Lean Six Sigma requires more than just following the basic steps. In this chapter, additional tips and suggestions will be provided to ensure a successful implementation of Lean Six Sigma in an organization.

Obtain support from top management

Successful implementation of Lean Six Sigma requires the support of top management in the organization. It is important for top management to understand the potential benefits of implementing Lean Six Sigma and be committed to the implementation process. This may involve allocating appropriate financial and human resources and appointing an experienced project leader.

Clearly communicate the implementation process

It is important for all relevant members of the organization to have a clear understanding of the Lean Six Sigma implementation process and be prepared for change. This may involve conducting information meetings and training sessions to explain the implementation process and potential advantages. It is also important to provide regular and transparent communication throughout the implementation process to keep organization members informed about progress and challenges.

Set achievable goals

Setting achievable and realistic goals for Lean Six Sigma implementation is important. The goals should be specific, measurable, achievable, relevant, and timely (SMART) and should reflect the challenges identified in the process assessment. It is also important to establish a detailed action plan to achieve these goals.

Involve all relevant members of the organization

Involving all relevant members of the organization in the Lean Six Sigma implementation process is crucial. This may involve forming project teams, conducting interviews, and gathering feedback from organization members to identify issues and challenges. It is also important to provide ongoing training and support to all organization members to ensure they are prepared for the change.

Regularly evaluate progress

Regularly evaluating progress towards established goals and adjusting the implementation plan as needed is important. This may involve conducting follow-up assessments and regularly reviewing improvement metrics to ensure that objectives are being achieved. It is also important to provide feedback and recognition to organization members who contribute to the implementation's success.

Benefits of Lean Six Sigma Implementation

Implementing Lean Six Sigma can provide numerous benefits to an organization. In this chapter, some of the key benefits of implementing Lean Six Sigma will be discussed.

Quality improvement

One of the main benefits of implementing Lean Six Sigma is the improvement in the quality of the organization's products and services. The Lean Six Sigma methodology focuses on defect elimination and reduction of variation in processes, which can significantly enhance the quality of products and services.

Cost reduction

Implementing Lean Six Sigma can also help an organization reduce costs by improving process efficiency and effectiveness. By eliminating inefficient processes and reducing process variation, an organization can lower production costs and improve profitability.

Increased productivity

Improving process efficiency and effectiveness can also lead to increased productivity. By reducing cycle times and improving the quality of products and services, an organization can enhance its ability to produce more in less time.

Enhanced customer satisfaction

Improving product and service quality and reducing cycle times can result in higher customer satisfaction. Customers can experience greater satisfaction by receiving higher-quality products and services in less time.

Organizational culture improvement

Implementing Lean Six Sigma can also improve an organization's culture. By involving all members of the organization in the continuous improvement process, a sense of ownership and responsibility for quality and process efficiency can be fostered. This can lead to a more positive and collaborative organizational culture.

Competitive advantage

Implementing Lean Six Sigma can provide a competitive advantage to an organization by improving quality, reducing costs, and increasing productivity. Organizations that implement Lean Six Sigma may have a competitive edge over those that do not by offering higher-quality products and services at more competitive prices.

Challenges of Lean Six Sigma Implementation

While implementing Lean Six Sigma can provide numerous benefits, it can also present challenges for an organization. In this chapter, some of the main challenges of Lean Six Sigma implementation will be discussed.

Resistance to change

One of the main challenges of Lean Six Sigma implementation is resistance to change from organization members. Implementing Lean Six Sigma may require significant changes in organizational culture and work processes, and some members of the organization may resist these changes.

Lack of top management commitment

Lack of commitment from top management can also be a challenge for Lean Six Sigma implementation. If top management is not fully committed to the implementation process, it can be difficult to obtain the necessary resources and support for successful implementation.

Lack of skills and knowledge

Lack of skills and knowledge in the Lean Six Sigma methodology can also be a challenge for implementation. It is important to provide appropriate training and support to all organization members to ensure they are prepared for the change.

Lack of financial and human resources

Lack of financial and human resources can also be a challenge for Lean Six Sigma implementation. Implementing Lean Six Sigma may require a significant investment in financial and human resources, and it can be difficult to obtain the necessary resources if the organization has limited resources.

Conclusion

Implementing Lean Six Sigma can provide numerous benefits to an organization, including quality improvement, cost reduction, increased productivity, enhanced customer satisfaction, and improved organizational culture. However, implementing Lean Six Sigma can also pose challenges, such as resistance to change, lack of top management commitment, lack of skills and knowledge, and lack of financial and human resources.

To ensure the success of Lean Six Sigma implementation, it is important to involve all members of the organization, provide appropriate training and support, and ensure top management commitment. It is also important to address challenges and issues as they arise to ensure the continuity of the continuous improvement process.

In summary, implementing Lean Six Sigma can be a challenging process, but the benefits can be significant for an organization. By focusing on continuous improvement and engaging all members of the organization, an organization can improve its quality, reduce costs, increase productivity, and provide higher customer satisfaction.

COMMUNICATION AND LEADERSHIP IN LEAN SIX SIGMA PROJECTS

Communication and leadership are two essential elements in any process improvement project, especially within the Lean Six Sigma framework. Effective communication is crucial to ensure that all stakeholders understand the project's objectives and expectations, as well as to maintain constant and transparent communication throughout the process. On the other hand, effective leadership is necessary to ensure that the project team is aligned around common goals and has the motivation and skills required to successfully complete the project. In this chapter, we will explore the relationship between communication and leadership in Lean Six Sigma projects and discuss some strategies and best practices to improve both elements within an organization's context.

Communication in Lean Six Sigma Projects

Communication is fundamental to the success of any project, but it is particularly important in Lean Six Sigma projects where objectives can be highly technical and processes can be complex. Effective communication is necessary to ensure that all stakeholders, including organizational leaders, the project team, and affected community members, understand the project's objectives and how it will be carried out. It is also important to maintain constant and transparent communication throughout the project's process.

One of the most effective ways to ensure effective communication in a Lean Six Sigma project is to establish a detailed and well-structured communication plan from the beginning. The communication plan should include a description of key audiences, key messages, and communication channels that will be used to reach each audience. It should also clearly establish the project's communication objectives and the schedule of communication activities to be conducted.

Additionally, it is important for project leaders and team members to communicate openly and regularly. This can include regular team meetings, regular status updates, and open and transparent communication about the challenges and opportunities that arise during the project.

Another effective strategy to improve communication in Lean Six Sigma projects is to use data visualization tools to communicate project results. Data visualization tools can help project team members and stakeholders understand project results more clearly and easily.

Leadership in Lean Six Sigma Projects

Effective leadership is crucial in any project, and it is particularly important in Lean Six Sigma projects where processes and objectives can be highly technical and complex. Project leaders must have the ability to motivate the project team and guide them effectively towards the project's goal.

One of the most effective ways to lead a Lean Six Sigma project is to establish a clear and shared vision of the project from the beginning. This can help align the project team around common project goals and ensure that all team members are motivated and working together effectively.

Another important strategy for leading Lean Six Sigma projects is to foster collaboration and teamwork. The Lean Six Sigma approach is based on the idea that project success depends on effective collaboration among all stakeholders. Project leaders must ensure that project team members work together effectively and that a culture of collaboration and mutual support is promoted.

Additionally, project leaders must have strong project management skills and be familiar with Lean Six Sigma tools and techniques. This can help ensure that the project is completed within the planned timeframe and budget, and that process improvement objectives are achieved.

Another important skill that project leaders must have is the ability to motivate the project team and maintain their commitment throughout the project process.

This can include recognizing the team's good work, providing regular feedback and support, and ensuring that team members have the necessary tools and resources to effectively complete the project.

Strategies to Improve Communication and Leadership in Lean Six Sigma Projects

There are several strategies and best practices that organizations can implement to improve both communication and leadership in Lean Six Sigma projects.

Establish a detailed communication plan: As mentioned earlier, establishing a detailed communication plan from the beginning of the project can help ensure effective and constant communication throughout the project process. The plan should include a description of key audiences, key messages, and communication channels that will be used to reach each audience.

Provide leadership training and development: Leadership training and development can help project leaders develop strong leadership skills and become familiar with Lean Six Sigma tools and techniques. This can improve their ability to lead and motivate the project team effectively.

Foster collaboration and teamwork: As mentioned earlier, fostering collaboration and teamwork is essential for the success of a Lean Six Sigma project. Project leaders must ensure that team members work together effectively and that a culture of collaboration and mutual support is promoted.

Use data visualization tools: Data visualization tools can help communicate project results more clearly and easily. Project leaders should consider using data visualization tools to effectively communicate project results.

Establish clear and shared goals: Establishing clear and shared goals from the beginning of the project can help align the project team around common project goals and maintain their motivation and commitment throughout the project process.

Conclusion

Communication and leadership are two essential elements in any process improvement project, especially within the Lean Six Sigma framework. Effective communication is fundamental to ensure that all stakeholders understand the project's objectives and expectations, as well as to maintain constant and open communication throughout the project process. On the other hand, effective leadership is key to ensuring that the project is completed within the planned timeframe and budget, and that process improvement objectives are achieved.

To improve communication and leadership in Lean Six Sigma projects, organizations should establish a detailed communication plan, provide leadership training and development, foster collaboration and teamwork, use data visualization tools, and establish clear and shared goals.

In summary, the success of any Lean Six Sigma project largely depends on effective communication and leadership. Project leaders must ensure that communication is effective and constant, and that a culture of collaboration and teamwork is promoted. Additionally, they must have strong leadership skills and be familiar with Lean Six Sigma tools and techniques. By implementing the best practices and strategies described in this chapter, organizations can improve communication and leadership in Lean Six Sigma projects and ultimately enhance processes and increase customer satisfaction.

IDENTIFICATION AND WASTE REDUCTION IN PROCESSES

In today's business world, cost reduction and process efficiency are key to a company's success. One way to achieve this is through the implementation of continuous improvement methodologies, such as Lean Six Sigma. In this chapter, we will discuss the identification and waste reduction in processes using the Lean Six Sigma methodology.

What is Lean Six Sigma?

Before diving into the details of how to identify and reduce waste in processes, it is important to understand what Lean Six Sigma is. Lean Six Sigma is a continuous improvement methodology that combines two approaches: Lean Manufacturing and Six Sigma.

Lean Manufacturing focuses on eliminating waste in processes, while Six Sigma focuses on reducing variability and improving quality. The combination of these two approaches results in a highly effective methodology for improving efficiency and quality in processes.

Identification of Waste in Processes

Before waste can be reduced in a process, it is important to identify it. Waste refers to any activity or process that does not add value to the final product or service. The following are the eight types of waste identified by Lean

Manufacturing:

Overproduction: Producing more than necessary or before it is needed.

Waiting Time: Time wasted waiting for a process or resource.

Transportation: Unnecessary movement of materials or products.

Overprocessing: Doing more than is needed to complete a task.

Inventory: Having more inventory than necessary.

Motion: Unnecessary movement of people or resources.

Defects: Errors in the process that result in defective products.

Underutilization of Talent: Not fully leveraging employees' potential.

To identify waste in a process, it is important to conduct a detailed analysis of the process and look for areas where the aforementioned wastes can be eliminated or reduced. This can be done through the creation of a value stream map, which shows the flow of the process from start to finish.

Once the wastes have been identified, measures can be taken to reduce or eliminate them entirely.

Waste Reduction in Processes

Once the wastes in a process have been identified, it is important to take measures to reduce or eliminate them. The following are some strategies that can be used to reduce waste in processes using the Lean Six Sigma methodology:

Implement Continuous Flow: This involves eliminating bottlenecks and waiting times in the process. By reducing waiting times and ensuring that processes are performed in sequence, the waste of transportation, motion, and overproduction can be reduced.

Reduce Variability: By reducing variability in the process, the waste of overprocessing and defects can be reduced. Tools such as Six Sigma can be used to identify and reduce variability in the process.

Implement the Pull System: This involves producing only what is needed, rather than overproducing and creating unnecessary inventory. By using the pull system, the waste of overproduction and inventory can be reduced.

Implement Kanban: This involves using visual signals to indicate when more of a product should be produced or when a product should be moved to the next stage of the process. By using kanban, the waste of inventory and transportation can be reduced.

Improve Process Efficiency: By improving process efficiency, the waste of motion and underutilization of talent can be reduced. Tools such as time and motion analysis can be used to identify areas where process efficiency can be improved.

Implement Kaizen Philosophy: This involves continuously improving the process through small improvements rather than big changes. By implementing the Kaizen philosophy, waste can be reduced in multiple areas of the process.

Train Employees: By training employees in Lean Six Sigma, they can be provided with the skills and tools necessary to identify and reduce waste in processes. This can lead to a culture of continuous improvement in the company.

Example of Lean Six Sigma Application in Waste Reduction

To illustrate how Lean Six Sigma can be applied to reduce waste in a process, let's use the example of a furniture manufacturing company. The company has identified that the furniture assembly process has significant waste due to overproduction and overprocessing.

To reduce the waste of overproduction, the company decides to implement the pull system and kanban. They use visual signals to indicate when more of a product is needed or when a product should be moved to the next stage of the process.

To reduce the waste of overprocessing, the company decides to use Six Sigma to reduce variability in the process. They use Six Sigma tools to identify areas where variability can be reduced and improve the processes to reduce overprocessing.

To ensure that the process is being carried out efficiently, the company uses time and motion analysis to identify areas where process efficiency can be improved. They also train employees in Lean Six Sigma to foster a culture of continuous improvement.

Conclusion

The identification and reduction of waste in processes are key to improving efficiency and reducing costs in a company. Lean Six Sigma is a highly effective methodology for achieving this. By identifying the waste in a process and taking measures to reduce or eliminate it, the efficiency and quality of processes in a company can be improved.

Furthermore, the implementation of Lean Six Sigma can lead to a culture of continuous improvement in the company, which can result in greater customer satisfaction and a competitive advantage in the market. It is important to note that the identification and reduction of waste in processes is not a one-time continuous process but rather an iterative process that requires constant improvement and monitoring.

IDENTIFICATION AND ELIMINATION OF PROCESS BOTTLENECKS

The identification and elimination of process bottlenecks is a key topic for improving the efficiency and profitability of a company. In the last decade, a highly effective methodology has been developed for achieving this: Lean Six Sigma. This approach combines the Lean methodology, which focuses on waste elimination, with the Six Sigma methodology, which focuses on reducing process variation. Together, these methodologies allow for the identification and elimination of process bottlenecks to improve quality, reduce costs, and increase customer satisfaction.

In this chapter, key concepts of Lean Six Sigma will be presented, and how this methodology can be applied to identify and eliminate process bottlenecks will be discussed. The tools and techniques used in Lean Six Sigma to achieve these objectives will also be discussed.

Key Concepts of Lean Six Sigma

To understand how Lean Six Sigma works, it is important to grasp the key concepts of both methodologies.

Lean Methodology

The Lean methodology originated in the Japanese automotive industry in the 1950s and focused on waste elimination in production processes. The five principles of the Lean methodology are:

Value: Identify what the customer considers valuable.

Value Stream: Identify the value stream in the process.

Flow: Create a continuous flow of work to eliminate waste.

Pull Production: Produce only what is needed, when it is needed.

Perfection: Strive for perfection in the process.

The Lean methodology focuses on eliminating or reducing eight types of waste, which are:

Overproduction, waiting, transportation, overprocessing, inventory, unnecessary motion, defects, underutilization of talent.

Six Sigma Methodology

The Six Sigma methodology was developed at Motorola in the 1980s and focused on reducing variation in production processes. Variation is defined as anything that can negatively impact the quality of a product or service.

The Six Sigma methodology is based on a strategy called DMAIC, which stands for:

Define: Define the problem or process to be improved.

Measure: Measure the current performance of the process.

Analyze: Analyze data to identify root causes of the problem.

Improve: Develop and implement solutions to address the root causes.

Control: Establish controls to sustain the improved process.

The goal of Six Sigma is to achieve a level of quality of 3.4 defects per million opportunities (DPMO). This means that the process produces high-quality products or services with a very low error rate.

Combining Lean and Six Sigma

Combining Lean and Six Sigma in the Lean Six Sigma methodology allows for leveraging the strengths of both methodologies to improve efficiency and quality in processes. By combining these two methodologies, it is possible to:

Identify and eliminate the eight types of waste identified by Lean.

Reduce variation in processes identified by Six Sigma.

Improve the quality of products or services.

Reduce costs and increase profitability.

Increase customer satisfaction.

The Lean Six Sigma methodology is based on five phases: Define, Measure, Analyze, Improve, and Control (DMAIC). Each phase has a specific set of activities and tools used to achieve the phase's objectives. The following describes each of the phases.

Define Phase

The first phase of the Lean Six Sigma methodology is the Define phase. In this phase, the problem or process to be improved is identified. The main activities in this phase include:

Identifying the problem or process to be improved.

Defining the project objective.

Forming a project team.

Developing a project plan.

To achieve these objectives, various tools and techniques are used, such as the project selection matrix, project charter, process map, and project scope definition.

Measure Phase

The second phase of the Lean Six Sigma methodology is the Measure phase. In this phase, the current performance of the process is measured, and areas for improvement are identified. The main activities in this phase include:

Identify process performance metrics.

Measure the current process performance.

Identify process improvement points.

To achieve these objectives, various tools and techniques are used, such as flowcharts, histograms, control charts, and process capability.

Analysis Phase

The third phase of the Lean Six Sigma methodology is the Analysis phase. In this phase, data is analyzed to identify the root causes of the problem. The main activities in this phase include:

Analyzing process data.

Identifying root causes of the problem.

Assessing the importance and feasibility of proposed solutions.

To achieve these objectives, various tools and techniques are used, such as Pareto analysis, root cause analysis, fishbone diagram, and prioritization matrix.

Improve Phase

The fourth phase of the Lean Six Sigma methodology is the Improve phase. In this phase, solutions are developed and implemented to address the root causes identified in the Analysis phase. The main activities in this phase include:

Generating potential solutions to address root causes.

Evaluating and selecting the best solutions.

Developing an implementation plan.

Implementing the solutions.

To achieve these objectives, various tools and techniques are used, such as design of experiments, cost-benefit analysis, pilot implementation, and control plan.

Control Phase

The fifth and final phase of the Lean Six Sigma methodology is the Control phase. In this phase, it is ensured that the improvements made in the Improve phase are sustained over time and the process is continuously improved. The main activities in this phase include:

Establishing process control and monitoring measures.

Implementing a control plan.

Training personnel and documenting the changes.

Conducting regular audits to evaluate process performance.

To achieve these objectives, various tools and techniques are used, such as control plan, trend analysis, personnel training, and documentation review.

Identification and Elimination of Bottlenecks

One of the main applications of Lean Six Sigma is the identification and elimination of bottlenecks in processes. A bottleneck is a stage in a process that limits the production capacity of the process. In other words, it is a point in the process where the production capacity is lower than the process demand.

Identifying and eliminating bottlenecks is essential for improving process efficiency and profitability. Eliminating a bottleneck increases the production capacity of the process and reduces waiting time, thereby improving service quality and customer satisfaction.

To identify and eliminate bottlenecks in a process, various tools and techniques are used, such as value-added analysis, capacity analysis, process flow analysis, and time and motion analysis.

Value-Added Analysis

Value-added analysis is a tool used to identify activities that add value in the process and activities that do not add value. The goal of value-added analysis is to eliminate activities that do not add value to improve process efficiency.

Value-added analysis is divided into three categories: value-added activities, necessary non-value-added activities, and unnecessary non-value-added activities. By eliminating unnecessary non-value-added activities, costs can be reduced and process efficiency can be improved.

Capacity Analysis

Capacity analysis is a tool used to evaluate the production capacity of a process. The goal of capacity analysis is to identify bottlenecks in the process and improve process efficiency.

Capacity analysis is divided into two categories: nominal capacity and actual capacity. Nominal capacity is the theoretical capacity of the process, while actual capacity is the real capacity of the process. By comparing nominal capacity and actual capacity, bottlenecks in the process can be identified.

Process Flow Analysis

Process flow analysis is a tool used to identify the stages in a process and the sequence of those stages. The goal of process flow analysis is to identify value-added stages and non-value-added stages to improve process efficiency.

Process flow analysis is divided into two categories: current process flow and ideal process flow. The current process flow describes how the process is currently performed, while the ideal process flow describes how the process should ideally be.

By comparing the current process flow and the ideal process flow, stages in the process that do not add value can be identified and eliminated to improve process efficiency.

Time and Motion Analysis

Time and motion analysis is a tool used to evaluate the time and movements required to perform a task in a process. The goal of time and motion analysis is to identify stages in the process that require more time and movements to perform the task and improve process efficiency.

Time and motion analysis is divided into two categories: cycle time and work time. Cycle time is the time required to complete a task, while work time is the time spent on the task. By reducing work time and cycle time, process efficiency can be improved.

Conclusion

The Lean Six Sigma methodology is an effective tool for identifying and eliminating bottlenecks in processes. The methodology is divided into five phases: Define, Measure, Analyze, Improve, and Control.

In the Define phase, the project scope is established, and project objectives and the project team are identified.

In the Measure phase, data is collected, and measures are established to evaluate process performance.

In the Analyze phase, the collected data is analyzed to identify problems in the process.

In the Improve phase, solutions are identified and implemented to improve the process.

In the Control phase, it is ensured that the improvements made in the Improve phase are sustained over time and the process is continuously improved.

To identify and eliminate bottlenecks in processes, various tools and techniques are used, such as value-added analysis, capacity analysis, process flow analysis, and time and motion analysis.

By identifying and eliminating bottlenecks in processes, process efficiency and profitability can be improved, resulting in increased customer satisfaction and improved service quality.

DESIGN OF EXPERIMENTS AND ANALYSIS OF VARIANCE

Design of Experiments (DOE) is a statistical technique used to investigate the relationship between process factors and their outcomes. DOE can be used to optimize a process, identify the root cause of a problem, and improve product or service quality. Lean Six Sigma methodology utilizes DOE as a key tool in its approach to continuous improvement. In this chapter, we will explain in detail how DOE is used with Lean Six Sigma and how Analysis of Variance (ANOVA) is performed to interpret experiment results.

Definition of Design of Experiments

DOE refers to a statistical technique used to determine how factors affect process outcomes. Factors can be input variables such as temperature, speed, pressure, time, etc. Outcomes can be anything measured in the process, such as product quality, cycle time, performance, etc. The goal of DOE is to identify which factors are significant and determine the best combination of factors to achieve desired outcomes.

Types of Design of Experiments

There are several types of DOE that can be used depending on the experiment's objective and process complexity. Some of the most common types include:

Single-variable design: Used to determine the effect of a single factor on the process outcome. One factor is varied at a time, and the outcome is measured.

Factorial design: Used to determine the effect of multiple factors and their interactions on the process outcome. Multiple factors are varied simultaneously, and the outcome is measured.

Taguchi design: Used to improve process robustness against variations in factors. Multiple factors are varied in specific combinations, and the outcome is measured.

Response surface design: Used to optimize input factors and find the combination that produces the best outcomes. Factors are varied within a range, and the outcome is measured at selected points within the range.

Steps for Design of Experiments with Lean Six Sigma

The process of Design of Experiments with Lean Six Sigma consists of the following steps:

Step 1: Define the experiment objective: In this step, the experiment objective is clearly defined. Desired outcomes are identified, and factors that can affect them are determined.

Step 2: Select the type of design of experiments: Once the experiment objective is defined, the type of DOE to be used is selected. Process complexity and the number of factors to be analyzed should be taken into account.

Step 3: Select factors and levels: In this step, the factors to be analyzed in the experiment are selected, and the levels for each factor are determined. It is important to select factors believed to be significant and have the greatest impact on the outcomes.

Step 4: Design the experiment: Once the factors and levels are selected, the experiment is designed. An experimental plan is established, including test and control groups, how factors will be varied, and how outcomes will be measured.

Step 5: Conduct the experiment: In this step, the experiment is conducted according to the experimental plan designed in the previous step. It is important to take accurate measurements and record data appropriately.

Step 6: Analyze the results: Once the data is collected, statistical analysis is performed to determine the relationship between factors and outcomes. ANOVA is used to determine if there are significant differences between test and control groups and to identify factors that have a significant impact on the outcomes.

Step 7: Interpret the results and take action: In this step, the results of the statistical analysis are interpreted, and actions are taken to improve the process. Lean Six Sigma tools such as Process Mapping and Root Cause Analysis can be used to identify areas for improvement and develop action plans for implementing those improvements.

Analysis of Variance (ANOVA)

ANOVA is a statistical technique used to compare the means of two or more groups and determine if there are significant differences among them. In the context of DOE, ANOVA is used to analyze the experiment results and determine if there are significant differences between test and control groups.

ANOVA is based on the null hypothesis that there are no significant differences among the groups, and the alternative hypothesis that there is at least one significant difference among the groups. ANOVA calculates the total sum of squares (SST), which represents the total variation in the data, and the sum of squares between groups (SSB), which represents the variation between the groups. The difference between SST and SSB is called the sum of squares within groups (SSW), which represents the variation within the groups. ANOVA also calculates the F-value, which is the ratio of variation between groups to variation within groups.

If the F-value is greater than the critical value, the null hypothesis is rejected, and it is concluded that there is at least one significant difference among the groups. In that case, multiple comparison tests can be performed to determine which groups differ significantly from each other.

Conclusions

Design of Experiments and Analysis of Variance are key statistical techniques used in Lean Six Sigma to improve product or service quality. DOE allows determining how factors affect process outcomes and finding the best combination of factors to achieve desired results. ANOVA is used to analyze experiment results and determine if there are significant differences between test and control groups. Both techniques are fundamental to the continuous improvement approach of Lean Six Sigma and valuable tools for any organization seeking to improve its quality and efficiency.

DEVELOPMENT OF INNOVATIVE SOLUTIONS USING LEAN SIX SIGMA

In the current business environment, innovation has become a necessity for companies that want to remain competitive in the market. The development of innovative solutions involves the application of methodologies and tools that enable companies to identify improvement opportunities and design efficient and effective solutions. Among these methodologies, Lean Six Sigma stands out, combining the principles of Lean Manufacturing and Six Sigma to improve the quality and efficiency of business processes. This chapter will explore how companies can use Lean Six Sigma to develop innovative solutions and how this methodology can help them stay competitive in the current market.

What is Lean Six Sigma?

Lean Six Sigma is a process improvement methodology that combines the principles of Lean Manufacturing and Six Sigma. Lean Manufacturing focuses on eliminating waste and improving process efficiency, while Six Sigma focuses on reducing variability and improving process quality. The combination of these two methodologies allows companies to improve the quality of their products and services, reduce costs, and enhance customer satisfaction.

The Lean Six Sigma methodology is divided into five phases: Define, Measure, Analyze, Improve, and Control. Each phase focuses on a specific part of the improvement process and utilizes specific tools and techniques to achieve the objectives of the phase. The ultimate goal of Lean Six Sigma is to create a more efficient and effective process that yields consistent and high-quality results.

Development of innovative solutions using Lean Six Sigma

The process of developing innovative solutions using Lean Six Sigma begins with defining the problem or improvement opportunity. This phase focuses on identifying the problem or improvement opportunity, establishing project objectives, and defining the project scope. Once the problem has been defined, the next phase is Measurement.

The Measurement phase focuses on collecting data about the current process and establishing a performance baseline. This baseline is used to measure the success of the project and determine if the proposed solutions have improved the process. Data collection tools such as control charts, flowcharts, and process maps are used to gather data.

The Analysis phase focuses on identifying the root causes of the problem or improvement opportunity. Tools such as Ishikawa diagrams, Pareto analysis, and value analysis are used to identify the underlying causes of the problem. Once the root causes have been identified, the Improvement phase follows.

The Improvement phase focuses on developing innovative solutions to address the root causes identified in the Analysis phase. The proposed solutions are tested and implemented on a small scale before being implemented across the entire process. Tools such as solution selection matrices and implementation plans are used to ensure that the proposed solutions are effective and efficient.

Finally, the Control phase focuses on maintaining and improving the process after the implementation of the solutions. Control measures are established, and process performance is monitored to ensure that the improvements are sustained over time. Tools used in this phase include control plans, control charts, and process audits.

Example of Lean Six Sigma application in developing innovative solutions

To illustrate the process of developing innovative solutions using Lean Six Sigma, let's consider an example of a car manufacturing company that wants to improve the efficiency of its assembly line.

Define: The company defines the problem as a high rework rate on the assembly line, resulting in decreased production and increased costs.

Measure: The company collects data on the current assembly process and establishes a performance baseline. A control chart is used to monitor the rework rate, and it is identified that the problem is concentrated in a specific stage of the process.

Analyze: The company uses an Ishikawa diagram to identify possible causes of the problem. It is discovered that one of the causes is an issue with the assembly tool used in that specific stage of the process.

Improve: The company develops an innovative solution that involves creating a more efficient and ergonomic assembly tool. The tool is tested on a small scale, and it is determined that it improves process efficiency and reduces the rework rate.

Control: The company implements the assembly tool across the entire assembly line and establishes control measures to monitor its performance. A control plan is used to ensure that the improved performance is maintained, and the process is monitored through control charts and process audits.

Results: The company achieves a significant reduction in the rework rate and an increase in production on the assembly line. Additionally, significant cost savings are achieved due to the reduction in the rework rate.

Conclusion

The use of Lean Six Sigma for the development of innovative solutions is a powerful tool for companies that want to stay competitive in the current market. This methodology allows companies to identify improvement opportunities, develop effective and efficient solutions, and sustain the improved performance over time. The process of developing innovative solutions using Lean Six Sigma is divided into five phases: Define, Measure, Analyze, Improve, and Control. Each phase utilizes specific tools and techniques to achieve the objectives of the phase and create a more efficient and effective process.

IDENTIFICATION AND ANALYSIS OF RISKS IN PROCESSES

Identification and analysis of risks are fundamental processes for any company seeking to improve its operations and reduce the risks associated with its processes. Currently, there are several methodologies used to identify and analyze risks, among which Lean Six Sigma stands out.

Lean Six Sigma is a methodology used to improve the quality of business processes. It is based on the idea that all companies have processes that contain errors and, therefore, can be improved to reduce costs, increase customer satisfaction, and enhance the quality of products and services. In this chapter, we will explore how Lean Six Sigma can be used to identify and analyze risks in business processes.

Risk Identification

Risk identification is the first step in the risk analysis process. The objective of risk identification is to identify all potential risks associated with a specific process. In this regard, Lean Six Sigma can be used to identify risks in business processes.

To identify risks in business processes, specific Lean Six Sigma tools such as process mapping, Pareto analysis, and cause-and-effect analysis need to be utilized. Process mapping is used to visualize the entire business process, including inputs, outputs, and subprocesses. Pareto analysis is used to identify the most common problems in a business process, while cause-and-effect analysis is used to identify the root causes of the problems.

Once the risks in the business process have been identified, a list of the identified risks must be created. This list should include a detailed description of the risk, as well as the potential impact of the risk on the business process. Additionally, risks should be classified according to their probability of occurrence and potential impact.

Risk Analysis

Once the risks in the business process have been identified, it is necessary to perform a risk analysis to determine the probability of occurrence and potential impact of each risk on the business process. Risk analysis is conducted using specific Lean Six Sigma tools such as Failure Mode and Effects Analysis (FMEA) and risk matrix.

FMEA is a tool used to identify and assess the potential effects of failures in a business process. It is employed to evaluate the probability of a failure occurring, the potential impact of the failure on the business process, and the probability of detecting the failure before it causes significant impact. The risk matrix is used to assess the probability of a risk occurring and its potential impact on the business process.

Once the FMEA and risk matrix analyses have been conducted, the identified risks must be prioritized based on their probability of occurrence and potential impact. The most critical risks should be addressed first to minimize their potential impact on the business process.

Risk Management

Once the risks in the business process have been identified and analyzed, measures need to be implemented to mitigate the identified risks. In this regard, Lean Six Sigma can be used to implement specific measures to mitigate identified risks.

To mitigate the identified risks, a risk management plan needs to be developed, which includes specific measures to mitigate the risks. This plan should encompass preventive measures, mitigation measures, and contingency measures.

Preventive measures are used to prevent the occurrence of identified risks. These measures are implemented before the risk occurs and may include staff training, process improvement, and the implementation of additional controls.

Mitigation measures are used to reduce the potential impact of identified risks. These measures are implemented after the risk has occurred and may include the implementation of contingency plans and alternative solutions.

Contingency measures are used to manage the identified risks in the event of their occurrence. These measures are implemented after the risk has occurred and may include the implementation of emergency plans and the assignment of specific responsibilities.

Monitoring and Control

Once risk management measures have been implemented, it is necessary to monitor and control the identified risks to ensure the effectiveness of the implemented measures. In this regard, Lean Six Sigma can be used to monitor and control the identified risks.

To monitor and control the identified risks, specific monitoring and control measures need to be implemented. These measures may include periodic review of identified risks, measurement of the potential impact of risks, and the implementation of additional measures if new risks are identified.

Additionally, it is necessary to establish a feedback process to ensure that identified risks are effectively managed. This feedback process should include periodic review of identified risks, identification of new mitigation measures if necessary, and the implementation of continuous improvement measures.

Conclusion

In conclusion, Lean Six Sigma is an effective methodology for identifying, analyzing, and managing risks in business processes. Risk identification is the first step in the risk analysis process and is used to identify all potential risks associated with a specific process. Once the risks in the business process have been identified, a risk analysis is performed to determine the probability of occurrence and potential impact of each risk on the business process.

Once the risks in the business process have been identified and analyzed, measures need to be implemented to mitigate the identified risks. These measures include preventive measures, mitigation measures, and contingency measures. It is necessary to monitor and control the identified risks to ensure the effectiveness of the implemented measures and establish a feedback process to ensure that identified risks are effectively managed.

VALUE CHAIN ANALYSIS AND DELIVERY TIME REDUCTION

In the business world, efficiency in managing production processes is a key factor in staying competitive in the market. One of the most popular methods for improving efficiency is Lean Six Sigma, which combines the principles of Lean Manufacturing and Six Sigma to eliminate waste and reduce variability in processes. Additionally, an important tool for improving efficiency is value chain analysis, which allows for the identification of key processes in production and points where delivery times can be reduced.

This chapter will analyze the application of Lean Six Sigma in reducing delivery times through value chain analysis. It will explain the methodology of Lean Six Sigma, how value chain analysis can be applied to identify critical processes, and the Lean Six Sigma tools that can be used to improve process efficiency.

Lean Six Sigma Methodology

Lean Six Sigma is a process improvement methodology that combines the principles of Lean Manufacturing and Six Sigma. Lean Manufacturing focuses on eliminating waste in processes, while Six Sigma focuses on reducing variability in processes. By combining these two methodologies, Lean Six Sigma aims to eliminate waste and reduce variability in processes to improve efficiency and product quality.

The Lean Six Sigma methodology consists of five phases:

Define: In this phase, the problem to be solved is defined, and project objectives are established.

Measure: Data is collected in this phase to assess the current performance of the process.

Analyze: The collected data is analyzed in this phase to identify weak points in the process and root causes of the problems identified in the previous phase.

Improve: Improvements are implemented in the process to reduce waste and variability.

Control: Controls are established in this phase to ensure that the implemented improvements are maintained in the long term.

Value Chain Analysis

Value chain analysis is a tool that allows for the identification of key processes in production and points where delivery times can be reduced. The value chain consists of all the processes necessary to produce a product, from the acquisition of raw materials to the delivery of the final product to the customer.

Value chain analysis can be divided into two phases:

Identification of key processes: In this phase, processes that are critical for production and have a significant impact on delivery times are identified.

Identification of improvement opportunities: In this phase, points where delivery times can be reduced through waste elimination and variability reduction are identified.

Lean Six Sigma Tools for Improving Efficiency

There are several Lean Six Sigma tools that can be used to improve process efficiency. Some of the most common tools include:

Value Stream Mapping: Value stream mapping is a tool that visualizes all the processes involved in the production of a product and the flows of information and materials between them. This tool is useful for identifying critical processes and points where delivery times can be reduced.

Kanban: Kanban is an inventory control system used to reduce inventory and delivery times. This system relies on the use of kanban cards that indicate when a product should be produced or when inventory should be replenished.

Poka-Yoke: Poka-Yoke is a technique used to prevent errors and reduce variability in processes. This technique involves incorporating devices or mechanisms that prevent errors from occurring in processes.

Root Cause Analysis: Root cause analysis is a technique used to identify the underlying causes of a problem in processes. This technique focuses on identifying the deeper causes of a problem rather than surface-level causes.

5S: 5S is a technique used to improve organization and cleanliness in processes. This technique involves implementing five steps: Sort, Set in Order, Shine, Standardize, and Sustain.

Example of Lean Six Sigma Application in Delivery Time Reduction

To illustrate the application of Lean Six Sigma in reducing delivery times, a hypothetical example of reducing delivery times in the production of automotive parts will be presented.

Define the problem: The problem to be solved is the long delivery time of automotive parts.

Measure: Data is collected on current delivery times, and processes with the greatest impact on delivery times are identified.

Analyze: The value stream mapping is used to identify critical processes, and root cause analysis is conducted to identify the causes of problems identified in critical processes.

Improve: Improvements are implemented in the identified critical processes from the analysis phase, such as implementing Poka-Yoke devices to reduce process variability and implementing the Kanban system to reduce inventory and delivery times.

Control: Controls are established to ensure that the implemented improvements are sustained in the long term, such as implementing the 5S technique to maintain organization and cleanliness in processes.

Conclusion

Value chain analysis and the Lean Six Sigma methodology are powerful tools for improving efficiency in production processes and reducing delivery times. The combination of these tools allows for the identification of critical processes in production and points where delivery times can be reduced through waste elimination and product quality improvement.

The application of these tools can generate significant benefits for companies, such as cost and delivery time reduction, improved product quality, error reduction, and enhanced customer satisfaction.

However, implementing these tools requires long-term commitment from the company and training employees in their application. Additionally, continuous monitoring and tracking of processes are important to ensure that improvements are sustained over time.

In summary, the combination of the Lean Six Sigma methodology and value chain analysis can be a powerful tool for improving production processes and reducing delivery times. The application of these tools can bring significant benefits to companies but requires long-term commitment and employee training in their application.

IMPLEMENTATION OF QUALITY MANAGEMENT SYSTEMS

Quality management is an essential tool for any organization seeking to improve the efficiency and effectiveness of its processes. In this context, Lean Six Sigma has become a popular methodology for implementing quality management systems in organizations of all types. This chapter will provide an overview of the Lean Six Sigma methodology and its application in the implementation of quality management systems.

What is Lean Six Sigma?

Lean Six Sigma is a process improvement methodology that combines two popular approaches: Lean and Six Sigma. Lean focuses on waste elimination and continuous improvement, while Six Sigma focuses on reducing variability and improving quality. The combination of these approaches allows organizations to maximize the efficiency and effectiveness of their processes.

Principles of Lean Six Sigma

The Lean Six Sigma methodology is based on a set of key principles that guide its application. These principles include:

Customer Focus

Customer focus is fundamental to Lean Six Sigma. This means understanding customer needs and designing processes that meet those needs.

Continuous Improvement

Continuous improvement is a key principle of Lean Six Sigma. This means that the organization constantly strives to improve its processes and reduce waste.

Variability Reduction

Six Sigma focuses on reducing variability in processes. This means that the organization seeks to minimize deviations from standard processes to improve quality and consistency.

Value-Centered Processes

The Lean Six Sigma methodology focuses on processes that generate value for the customer. This means identifying and improving critical processes that directly impact customer satisfaction.

Data and Fact-Driven

Lean Six Sigma relies on data and facts to make informed decisions. This means collecting and analyzing data to identify root causes of problems and taking action to resolve them.

Teamwork Focus

Teamwork is essential for the implementation of Lean Six Sigma. This means involving all members of the organization in the process of continuous improvement and fostering collaboration and communication among them.

Benefits of Lean Six Sigma

The Lean Six Sigma methodology offers several benefits to organizations that implement it. Some of these benefits include:

Improved Efficiency

Lean Six Sigma enables organizations to improve the efficiency of their processes by eliminating waste and reducing variability.

Improved Quality

Six Sigma focuses on improving process quality. This means reducing defects and improving the consistency of products and services.

Increased Customer Satisfaction

Implementing Lean Six Sigma allows organizations to improve the quality of their products and services, which in turn can increase customer satisfaction. By effectively meeting customer needs, the organization can maintain customer loyalty and commitment.

Cost Reduction

Waste elimination and improved efficiency can help reduce costs for the organization. This can lead to increased profitability and competitiveness.

Strengthened Organizational Culture

Implementing Lean Six Sigma can help strengthen the organizational culture by fostering collaboration, teamwork, and continuous improvement at all levels of the organization.

Implementing Quality Management Systems with Lean Six Sigma

The implementation of quality management systems with Lean Six Sigma focuses on process improvement and waste elimination to enhance quality and efficiency. By combining the Lean Six Sigma methodology with quality management principles, organizations can significantly improve their processes and deliver high-quality products and services.

Planning

The first step in implementing a quality management system with Lean Six Sigma is planning. This involves defining project goals and establishing a detailed plan for implementation. During this phase, the necessary resources are identified, and key performance indicators (KPIs) are established to measure project success.

Analysis

In the analysis phase, data is collected and analyzed to identify problems and improvement opportunities. This involves using Lean Six Sigma tools and techniques to gather and analyze data and determine the root causes of problems. Some common tools used in this phase include value stream mapping, Pareto analysis, and the Ishikawa diagram.

Design

In the design phase, solutions are developed to address the problems identified in the analysis phase. This involves designing new processes and procedures that reduce waste and improve quality. During this phase, testing and simulations are also conducted to ensure that the proposed solutions are effective and efficient.

Implementation

In the implementation phase, the solutions developed in the design phase are put into practice. This involves training personnel and implementing new processes and procedures. During this phase, the necessary controls are also established to ensure process consistency and quality.

Control

In the control phase, processes are monitored and measured to ensure that the quality and efficiency levels established in the design phase are maintained. This involves implementing monitoring and feedback systems to detect problems and improvement opportunities. Action plans are also established to address any issues that arise during this phase.

Challenges in Implementing Quality Management Systems with Lean Six Sigma

While implementing quality management systems with Lean Six Sigma can offer many benefits, there are also challenges that can arise during the implementation process. Some common challenges in implementing quality management systems with Lean Six Sigma are described below:

Resistance to Change

Resistance to change is a common challenge in implementing quality management systems with Lean Six Sigma. Employees may be hesitant to change how tasks are performed and adopt new processes and procedures. To overcome this challenge, it is important to clearly communicate the benefits of the quality management system and provide training and support to employees throughout the implementation process.

Lack of Leadership and Commitment

Lack of leadership and commitment is another common challenge in implementing quality management systems with Lean Six Sigma. If organizational leaders are not committed to the implementation process, it is likely that employees will also lack commitment. To overcome this challenge, it is important for organizational leaders to actively support and promote the quality management system.

Lack of Resources

Implementing a quality management system with Lean Six Sigma may require significant resources, such as time, money, and trained personnel. If the organization lacks sufficient resources to effectively implement the quality management system, the process may not succeed. To overcome this challenge, it is important to allocate the appropriate resources and ensure they are used effectively.

Lack of Alignment with Organizational Strategy

If the quality management system is not aligned with the overall strategy of the organization, desired results may not be achieved. It is important for the implementation of the quality management system to be aligned with the organization's overall strategy, and clear objectives should be established to measure the success of the system.

Conclusions

Implementing quality management systems with Lean Six Sigma can help organizations improve the quality and efficiency of their processes, leading to increased customer satisfaction, profitability, and a stronger organizational culture. However, implementing quality management systems with Lean Six Sigma can also present challenges, such as resistance to change, lack of leadership and commitment, lack of resources, and lack of alignment with the organization's strategy.

To overcome these challenges, careful planning of the quality management system implementation is important. Clear communication of the system's benefits to employees, allocation of appropriate resources, and ensuring alignment with the organization's overall strategy are key. With the right approach, implementing quality management systems with Lean Six Sigma can be an effective and beneficial process for the organization.

INTEGRATION OF LEAN SIX SIGMA WITH OTHER PROCESS IMPROVEMENT METHODOLOGIES

In this chapter, we will explore how to integrate Lean Six Sigma with other process improvement methodologies such as Total Quality Management (TQM), Business Process Reengineering (BPR), Kaizen, and Agile. We will analyze the key concepts of each methodology and how they can be integrated to achieve effective improvement results.

Total Quality Management (TQM)

Total Quality Management (TQM) is a methodology that focuses on continuous improvement of quality through customer satisfaction, waste elimination, and improvement of business processes. TQM is based on the principle that quality is the responsibility of everyone in the organization and that continuous improvement should be sought in all aspects of the business process.

To integrate Lean Six Sigma with TQM, it is important to understand how the methodologies overlap and how they can complement each other. Both methodologies focus on continuous process improvement and waste elimination. However, Lean Six Sigma focuses more on reducing variability and improving efficiency, while TQM focuses on customer satisfaction and quality improvement.

One way to integrate Lean Six Sigma with TQM is to use Lean Six Sigma tools to identify areas for improvement in the process and then use TQM principles to implement the necessary changes to improve quality and customer satisfaction.

Additionally, the TQM methodology can help ensure that the implemented changes are sustainable in the long term by involving everyone in the organization in the process of continuous improvement.

Business Process Reengineering (BPR)

Business Process Reengineering (BPR) is a methodology that focuses on the radical restructuring of business processes to achieve significant improvements in efficiency and quality. BPR focuses on eliminating non-value-added activities, simplifying processes, and removing organizational barriers.

To integrate Lean Six Sigma with BPR, it is important to understand how the methodologies overlap and how they can complement each other. Both methodologies focus on improving efficiency and eliminating waste. However, BPR focuses on the radical restructuring of business processes, while Lean Six Sigma focuses on reducing variability and improving efficiency.

One way to integrate Lean Six Sigma with BPR is to use Lean Six Sigma tools to identify processes that require radical restructuring and then apply BPR principles to make significant changes to the process. Additionally, the Lean Six Sigma methodology can help ensure that the implemented changes are effective and sustainable in the long term by continuously monitoring and measuring process performance.

Kaizen

Kaizen is a Japanese methodology for continuous improvement that focuses on the gradual and constant improvement of business processes through the involvement of all members of the organization. Kaizen is based on the principle that small continuous improvements in the process can add up to significant improvements in the final outcome.

To integrate Lean Six Sigma with Kaizen, it is important to understand how the methodologies overlap and how they can complement each other. Both methodologies focus on continuous improvement and waste elimination. However, Kaizen focuses on gradual and constant improvement, while Lean Six Sigma focuses on reducing variability and improving efficiency.

One way to integrate Lean Six Sigma with Kaizen is to use Lean Six Sigma tools to identify areas for improvement in the process and then apply Kaizen principles to make gradual and constant improvements to the process. Additionally, the Kaizen methodology can help ensure that all members of the organization are involved in the process of continuous improvement and that improvements are made in all aspects of the business process.

Agile

Agile is a project management methodology that focuses on the rapid and continuous delivery of value to the customer through collaboration and constant adaptation. Agile is based on the principle that project requirements and objectives can change over time and, therefore, should be continuously adapted to meet customer needs.

To integrate Lean Six Sigma with Agile, it is important to understand how the methodologies overlap and how they can complement each other. Both methodologies focus on continuous improvement and waste elimination. However, Agile focuses on the rapid and continuous delivery of value to the customer, while Lean Six Sigma focuses on reducing variability and improving efficiency.

One way to integrate Lean Six Sigma with Agile is to use Lean Six Sigma tools to identify areas for improvement in the process and then apply Agile principles to continuously adapt and improve the process to meet customer needs. Additionally, the Agile methodology can help ensure that the improvement approach remains aligned with changing customer needs and that value is delivered quickly and effectively.

Conclusion

The integration of Lean Six Sigma with other process improvement methodologies can help achieve a holistic approach to business process improvement. By using the tools and principles of multiple methodologies, significant improvements in efficiency, quality, and customer satisfaction can be achieved. It is important to understand how the methodologies overlap and complement each other and how they can be applied together to achieve the best results. By integrating Lean Six Sigma with TQM, BPR, Kaizen, and Agile, continuous and sustainable improvement in the business process can be achieved.

It is important to remember that the integration of these methodologies is not a one-size-fits-all approach. Selecting the appropriate methodology for a specific situation should be based on a deep understanding of the business processes, challenges, and customer needs. It is essential to involve all members of the organization in the process of continuous improvement and establish a collaborative approach to achieve the best results.

In summary, integrating Lean Six Sigma with other process improvement methodologies can provide a holistic and effective approach to continuous improvement of the business process. Selecting the appropriate methodology should be based on a deep understanding of the business processes and customer needs. By involving all members of the organization and establishing a collaborative approach, significant and sustainable improvements in efficiency, quality, and customer satisfaction can be achieved.

PROBLEM-SOLVING COMPLEX PROBLEMS USING LEAN SIX SIGMA

Lean Six Sigma is a methodology that has evolved over the years from two different approaches: Lean Manufacturing and Six Sigma. The Lean Manufacturing approach originated in Japan in the 1950s and focused on waste elimination and continuous improvement. On the other hand, Six Sigma originated at Motorola in the 1980s and focused on quality improvement through variation reduction. The combination of these two approaches resulted in the Lean Six Sigma methodology, which has been used in a wide variety of industries, including manufacturing, healthcare, and financial services.

Basic principles of Lean Six Sigma

The basic principles of Lean Six Sigma are based on continuous improvement and waste elimination. These principles are applied through five phases: Define, Measure, Analyze, Improve, and Control (DMAIC).

Define

The first phase of DMAIC involves defining the problem to be solved and establishing project objectives. In this phase, it is important to identify stakeholders and establish a common understanding of the problem and project objectives.

Measure

The second phase of DMAIC involves measuring the current process and collecting relevant data. In this phase, tools such as process maps and control charts can be used to analyze the data and establish a baseline for the project.

Analyze

The third phase of DMAIC involves analyzing the collected data and determining the root causes of the problem. In this phase, tools such as Pareto analysis and cause-and-effect analysis are used to identify the fundamental causes of the problem.

Improve

The fourth phase of DMAIC involves improving the process through solution implementation and pilot testing. In this phase, it is important to ensure that the implemented solutions are sustainable in the long term and do not create new problems.

Control

The fifth phase of DMAIC involves controlling the improved process and monitoring its performance over time. In this phase, tools such as control charts and control plans are used to maintain the process in its new and improved condition.

Lean Six Sigma approach in solving complex problems

The Lean Six Sigma methodology focuses on solving complex problems through continuous improvement. The methodology focuses on identifying and eliminating waste in processes, which leads to improvement in business quality, efficiency, and profitability. Continuous improvement is an iterative process that involves problem identification and solution implementation to resolve them.

Continuous improvement is achieved through the implementation of the DMAIC methodology of Lean Six Sigma. This methodology provides a structured framework for problem-solving, enabling better understanding of the problem and more effective implementation of solutions. Additionally, the methodology encourages collaboration among stakeholders and collection of relevant data to make informed decisions.

The Lean Six Sigma methodology also emphasizes the elimination of waste in processes. Waste refers to activities that do not add value to the process and consume unnecessary resources. By eliminating these wastes, process efficiency improves, cycle time reduces, which in turn increases customer satisfaction and reduces production costs.

Furthermore, the Lean Six Sigma methodology emphasizes the importance of continuous improvement over time. Implementing solutions alone is not sufficient to permanently solve a problem. It is important to monitor the process performance and make adjustments as necessary to maintain continuous improvement.

In summary, the Lean Six Sigma approach to solving complex problems focuses on continuous improvement and waste elimination. The DMAIC methodology provides a structured framework for problem-solving, while waste elimination improves process efficiency and profitability. Continuous improvement over time is key to maintaining the benefits of solution implementation and ensuring customer satisfaction.

Application of Lean Six Sigma in practice

In this chapter, we will explore how Lean Six Sigma is applied in practice to solve complex problems in different industries. Case studies will be presented to illustrate how the Lean Six Sigma methodology can be used to solve complex problems and improve process efficiency and quality.

Application of Lean Six Sigma in the manufacturing industry

In the manufacturing industry, Lean Six Sigma is used to improve process efficiency and reduce production costs. An example of this is the implementation of Lean Six Sigma in an electronic component manufacturing company. The company faced quality issues and product rejections, resulting in increased production costs.

To address this problem, the DMAIC methodology of Lean Six Sigma was used to identify the root causes of the problem and develop effective solutions. Pilot testing was conducted to validate the solutions before implementation, ensuring long-term effectiveness of the solutions.

The implementation of Lean Six Sigma in this company resulted in a 40%

reduction in product rejections and significant improvement in product quality. Additionally, process efficiency improved, resulting in reduced production costs.

Application of Lean Six Sigma in the service industry

In the service industry, Lean Six Sigma is used to improve service quality and reduce waiting times. An example of this is the implementation of Lean Six Sigma in a hospital to improve patient care efficiency.

The hospital faced issues with patient waiting times and lack of coordination between departments. The DMAIC methodology of Lean Six Sigma was used to identify the root causes of the problem and develop effective solutions. Changes were implemented in appointment scheduling and task assignment to improve coordination between departments.

The implementation of Lean Six Sigma in this hospital resulted in a significant reduction in patient waiting times and improved coordination between departments. Additionally, patient satisfaction improved, leading to an enhancement of the hospital's reputation.

Application of Lean Six Sigma in the technology industry

In the technology industry, Lean Six Sigma is used to improve efficiency in product development and reduce delivery times. An example of this is the implementation of Lean Six Sigma in a software company to improve the software development process.

The company faced issues with software delivery times and lack of quality in the final product. The DMAIC methodology of Lean Six Sigma was used to identify the root causes of the problem and develop effective solutions. Changes were implemented in the software development process to improve product quality and reduce delivery times.

The implementation of Lean Six Sigma in this company resulted in a significant improvement in product quality and a reduction in software delivery times. Additionally, process efficiency improved, resulting in reduced production costs.

Application of Lean Six Sigma in the construction industry

In the construction industry, Lean Six Sigma is used to improve construction process efficiency and reduce construction times. An example of this is the implementation of Lean Six Sigma in an office building construction project.

The project faced issues with construction times and lack of coordination between construction teams. The DMAIC methodology of Lean Six Sigma was used to identify the root causes of the problem and develop effective solutions. Changes were implemented in the construction scheduling and coordination between teams to improve process efficiency.

The implementation of Lean Six Sigma in this project resulted in a significant reduction in construction times for the office building and improved coordination between construction teams. Additionally, the quality of the final building improved, leading to enhanced customer satisfaction.

In summary, the application of Lean Six Sigma in different industries demonstrates its effectiveness in solving complex problems and improving process efficiency and quality. The DMAIC methodology provides a structured framework for problem identification and effective solution implementation. Continuous improvement is key to maintaining the benefits of solution implementation and ensuring customer satisfaction.

LEAN SIX SIGMA PROJECT MANAGEMENT

Lean Six Sigma project management can be applied to any type of project, from manufacturing to customer service, and can be used by any organization seeking to improve its efficiency and effectiveness. In this chapter, the Lean Six Sigma project management methodology, its tools and techniques, and how it can be applied in different contexts will be described.

Lean Six Sigma: A Brief Introduction Lean Six Sigma combines two approaches: Lean and Six Sigma. Lean focuses on eliminating unnecessary processes and reducing delivery time, while Six Sigma focuses on improving the quality of products and services. The combination of both approaches allows organizations to improve the quality of their products and services while reducing costs and increasing efficiency.

The Lean Six Sigma methodology is divided into five phases: Define, Measure, Analyze, Improve, and Control (DMAIC). Each phase has a series of associated activities and tools that are used to identify issues, measure current performance, analyze data, identify improvement opportunities, implement solutions, and control long-term performance.

The Define phase focuses on defining the problem and establishing the project scope. In this phase, project goals are established, the project team is identified, and a project plan is established.

The Measure phase focuses on measuring current performance and establishing a baseline. In this phase, the data necessary to measure current performance is identified, and this data is collected and analyzed.

The Analyze phase focuses on analyzing the data collected in the previous phase. In this phase, the root causes of identified problems are identified, and improvement opportunities are identified.

The Improve phase focuses on implementing solutions to improve performance. In this phase, potential solutions are developed, and the most suitable solution is selected. Then the solution is implemented, and performance is tracked.

The Control phase focuses on controlling long-term performance. In this phase, controls are established to ensure that the implemented solution remains effective, and data continues to be collected and analyzed.

Tools and Techniques of the Lean Six Sigma Methodology The Lean Six Sigma methodology uses a series of tools and techniques to assist in each phase of the project. Some of the most common tools include:

Pareto Analysis: This tool is used to identify the most significant problems. Pareto analysis is based on the principle that 80% of problems are due to 20% of causes.

Ishikawa Diagram: Also known as a fishbone diagram or cause-and-effect diagram, this tool is used to identify the root causes of a problem. The diagram is divided into categories, and arrows are used to show the relationship between causes and the problem.

Value Stream Mapping: This tool is used to visualize the flow of the process and identify improvement points. The value stream map shows all the steps necessary to deliver a product or service, from raw material to the end customer.

Control Chart: This tool is used to monitor process performance over time. The control chart shows the upper and lower control limits, and process data is plotted over time.

Process Capability Analysis: This tool is used to measure the process's ability to produce products or services within specifications. It is used to determine if the process is operating within specifications and if improvements are required.

Regression Analysis: This tool is used to identify the relationship between two variables. It is used to determine if one variable affects the outcome of another variable and to what extent.

Application of the Lean Six Sigma Methodology in Different Contexts

The Lean Six Sigma methodology can be applied in a wide range of contexts, from manufacturing to customer service. Below are some examples of how the Lean Six Sigma methodology can be applied in different contexts.

Manufacturing

In the context of manufacturing, the Lean Six Sigma methodology can be used to improve efficiency and reduce production costs. For example, the methodology can be used to identify bottlenecks in the production line and improve process efficiency. It can also be used to reduce production cycle time and improve product quality.

Financial Services

In the context of financial services, the Lean Six Sigma methodology can be used to improve process efficiency and reduce costs. For example, it can be used to identify unnecessary processes in mortgage loan management and eliminate them to reduce processing costs. It can also be used to improve the quality of customer service by reducing response time and increasing customer satisfaction.

Healthcare

In the context of healthcare, the Lean Six Sigma methodology can be used to improve the quality of patient care and reduce healthcare costs. For example, it can be used to reduce waiting time in emergency rooms and improve the efficiency of admission and discharge processes. It can also be used to improve patient safety and reduce medical errors.

Conclusions

The Lean Six Sigma project management methodology is a powerful tool for improving efficiency and quality in any type of organization. By combining Lean and Six Sigma approaches, unnecessary processes can be identified and eliminated, and the quality of products and services can be improved. The methodology is divided into five phases: Define, Measure, Analyze, Improve, and Control (DMAIC), and uses a series of tools and techniques to assist in each project phase. The Lean Six Sigma methodology can be applied in a wide range of contexts, from manufacturing to healthcare and financial services.

One of the main advantages of the Lean Six Sigma methodology is its focus on continuous improvement. This means that projects are never considered "finished," but are constantly reviewed and improved to ensure that processes remain efficient and of high quality. Furthermore, by focusing on data and measurement, the methodology helps organizations make fact-based decisions rather than assumptions.

However, successful implementation of the Lean Six Sigma methodology requires proper planning and training. It is important for organizations to understand the principles and tools of the methodology and to train their personnel in their use. It is also essential for organizations to allocate sufficient time and resources to the project and to have the support of top management to ensure its success.

In summary, the Lean Six Sigma methodology is a powerful tool for improving efficiency and quality in any type of organization. By combining Lean and Six Sigma approaches, unnecessary processes can be identified and eliminated, and the quality of products and services can be improved. The methodology is divided into five phases: Define, Measure, Analyze, Improve, and Control (DMAIC), and uses a series of tools and techniques to assist in each project phase. The Lean Six Sigma methodology can be applied in a wide range of contexts and focuses on continuous improvement through data and measurement. Successful implementation requires planning, training, and support from top management.

DEVELOPMENT AND MANAGEMENT OF LEAN SIX SIGMA TEAMS

The development and management of Lean Six Sigma teams is a critical process for any organization seeking to improve its processes and increase customer satisfaction. In this chapter, we will discuss key aspects of Lean Six Sigma team development and management, including team formation, change management, effective communication, and performance measurement.

Team Formation

Team formation is the first critical step in developing an effective Lean Six Sigma team. Lean Six Sigma teams typically consist of individuals from different departments and functions, and can include both frontline employees and management-level employees. It is important for team members to have complementary skills and be willing to work together to achieve project objectives.

Team formation generally begins with the selection of team members. This involves identifying employees who have specific skills and knowledge in the project focus area. For example, if the project focuses on improving supply chain efficiency, team members with experience in logistics, procurement, and production may be selected.

Once the team has been selected, it is important to provide them with the necessary training to understand the basic concepts of Lean Six Sigma and apply them to the project. This may include training courses in Lean Six Sigma tools, such as Ishikawa diagrams and value stream maps, as well as training in soft skills such as leadership and communication.

Change Management

Change management is another critical aspect of Lean Six Sigma team development and management. Lean Six Sigma projects often involve significant changes to an organization's processes, technology, and culture. Resistance to change is common and can be a major obstacle to project success.

To effectively manage change, it is important to clearly communicate the project goals and potential benefits of implementing Lean Six Sigma. This may include cost reduction, quality improvement, and customer satisfaction. It is important to involve all employees affected by the project and provide them with the necessary training and resources to contribute to the project's success.

Effective Communication

Effective communication is another critical factor in Lean Six Sigma team development and management. Effective communication is important to ensure that all team members are aligned with project objectives, understand their roles and responsibilities, and are willing to work together to achieve the objectives.

Effective communication is also important to keep all stakeholders informed about the project's progress. This may include regular progress reports, email updates, and regular team meetings. It is important for communication to be clear and concise and tailored to the appropriate audience. For example, different communication may be needed for frontline employees and senior executives.

Performance Measurement

Performance measurement is a critical aspect of Lean Six Sigma team development and management. It is important to effectively measure project performance and results to determine if project objectives have been achieved and to identify areas for improvement for future projects.

Performance measurement generally involves tracking key project metrics such as cycle time, cost, quality, and customer satisfaction. These metrics should be measurable and quantifiable, and specific goals should be established for each of them.

It is important to use Lean Six Sigma tools such as statistical process control (SPC) to measure and analyze data. This can help identify patterns and trends and enable data-driven decision-making.

In addition to measuring project performance, it is also important to measure team performance. This may include measuring team satisfaction, team effectiveness, and individual skill development.

Conclusion

In summary, the development and management of Lean Six Sigma teams are critical to the success of any process improvement project. Effective team formation, change management, effective communication, and performance measurement are key aspects that must be considered to achieve project objectives and improve customer satisfaction. It is important for organizations to allocate the necessary resources to develop effective Lean Six Sigma teams and support them as they implement process improvement projects.

TRAINING AND CERTIFICATION OF LEAN SIX SIGMA PROFESSIONALS

To effectively implement Lean Six Sigma, it is necessary to have trained and certified professionals in the methodology. In this chapter, we will explore the process of training and certifying professionals in Lean Six Sigma, from the basics to the advanced levels of certification.

Basics of Lean Six Sigma

Before delving into the training and certification process, it is important to understand the basics of Lean Six Sigma. Below is a brief introduction to some of the most important terms:

Waste: Any activity that does not add value to a process is considered waste. The seven most common types of waste in processes are overproduction, waiting time, unnecessary transportation, unnecessary processing, unnecessary inventory, unnecessary motion, and defects.

Variability: Variability refers to any fluctuation in the process that can negatively affect quality or efficiency. Reducing variability is one of the main goals of the Six Sigma methodology.

DMAIC: DMAIC is the primary methodology used in Lean Six Sigma. It stands for Define, Measure, Analyze, Improve, and Control. Each stage of the process focuses on a different aspect of continuous improvement.

Black Belt: A Black Belt is a certified professional in Lean Six Sigma who has completed advanced level training and has the ability to lead complex projects of continuous improvement.

Green Belt: A Green Belt is a certified professional in Lean Six Sigma who has completed intermediate level training and has the ability to lead simple projects of continuous improvement.

Levels of Certification in Lean Six Sigma

There are several levels of certification in Lean Six Sigma, each with different requirements and necessary skills. Below are the most common certification levels:

Yellow Belt: A Yellow Belt is a professional who has completed basic training in Lean Six Sigma. Yellow Belts have a basic understanding of the methodology and can contribute to simple projects of continuous improvement.

Green Belt: As mentioned earlier, a Green Belt is a professional who has completed intermediate level training in Lean Six Sigma. Green Belts have a deeper understanding of the methodology and can lead simple projects of continuous improvement.

Black Belt: A Black Belt is a professional who has completed advanced level training in Lean Six Sigma. Black Belts have a comprehensive understanding of the methodology and can lead complex projects of continuous improvement.

Master Black Belt: A Master Black Belt is a professional who has completed an even more advanced level of training in Lean Six Sigma. Master Black Belts have a deep knowledge of the methodology and can lead multiple projects and train other professionals in Lean Six Sigma.

Champion: A Champion is a leader in the organization who supports and promotes the implementation of Lean Six Sigma throughout the company. Champions are not necessarily certified in Lean Six Sigma, but they have an important role in the success of the methodology in the organization.

Each certification level has different requirements in terms of training hours, practical experience, and skills needed to be certified. Additionally, each certification level has different responsibilities and expectations in terms of leading continuous improvement projects.

Training Process in Lean Six Sigma

The training process in Lean Six Sigma varies depending on the training provider and the organization seeking certification. However, in general, the training process in Lean Six Sigma consists of the following steps:

Identification of training needs: The organization must assess whether it needs training in Lean Six Sigma and which level of certification is necessary to meet its continuous improvement goals.

Selection of training provider: The organization must select a trusted training provider that offers Lean Six Sigma training courses at the required level.

Training: Professionals seeking certification must complete the required training hours and acquire the necessary skills to meet the requirements of the desired certification level.

Practical project: Professionals must complete a practical project of continuous improvement that demonstrates their ability to apply the methodology in a real-world setting.

Certification exam: Professionals must pass a certification exam that assesses their knowledge and skills in Lean Six Sigma.

Certification: Once all requirements are met, the professional receives certification in Lean Six Sigma at the desired level.

It is important to note that the training and certification process in Lean Six Sigma is a long-term commitment and requires significant dedication of time and resources. However, the benefits of the methodology are equally significant and can have a positive impact throughout the organization.

Benefits of Lean Six Sigma Certification

Lean Six Sigma certification can provide several benefits for professionals and the organization as a whole. Below are some of the most common benefits:

Improved efficiency: Lean Six Sigma methodology focuses on waste elimination and process optimization, which can result in increased efficiency in all areas of the organization.

Error reduction: By focusing on the elimination of variations and errors, Lean Six Sigma can help reduce the number of errors in processes, improving the quality of products and services and reducing costs associated with error correction.

Improved customer satisfaction: By improving the quality of products and services, Lean Six Sigma can help enhance customer satisfaction and brand loyalty.

Cost savings: Improved efficiency and error reduction can lead to a reduction in operating costs and improved profitability.

Improved organizational culture: By adopting Lean Six Sigma as a continuous improvement methodology, organizations can create a culture of continuous improvement that fosters innovation and collaboration.

Professional development: Lean Six Sigma certification can be a valuable asset in an individual's professional development, enhancing their career prospects and growth opportunities within the organization.

Challenges of Lean Six Sigma Implementation

Despite the potential benefits of Lean Six Sigma, implementing the methodology can present various challenges for organizations. Below are some of the common challenges:

Resistance to change: Implementing Lean Six Sigma may require significant changes in organizational culture and existing processes, which can lead to resistance to change from employees and leaders.

Lack of top management commitment: Effective implementation of Lean Six Sigma requires commitment and support from top management. If leaders are not committed to the methodology, there may be difficulties in implementing it effectively.

Lack of resources: Implementing Lean Six Sigma requires significant resources in terms of time, money, and trained personnel. If an organization is unwilling or lacks the resources to invest in implementing the methodology, it may be challenging to achieve the potential benefits.

Lack of understanding: If employees do not understand the methodology or its importance to the organization, there may be difficulties in implementing it effectively. Education and training are crucial to ensure that all employees understand and support the implementation of Lean Six Sigma.

Difficulties in measuring success: Implementing Lean Six Sigma requires constant measurement and evaluation of processes to identify improvement opportunities. If an organization has difficulties in measuring success or setting clear goals, it may be challenging to implement the methodology effectively.

Conclusion

Implementing Lean Six Sigma can be a valuable tool for improving efficiency, reducing errors, enhancing quality, and increasing customer satisfaction. However, the successful implementation of the methodology requires a significant commitment of time, resources, and leadership to overcome the challenges that may arise during the implementation process.

Training and certification in Lean Six Sigma can help professionals acquire the necessary skills to lead continuous improvement projects and provide a valuable asset in their professional development. Additionally, education and training are crucial to ensure that all employees understand the methodology and support its implementation.

It is important to note that Lean Six Sigma is not a one-size-fits-all solution for all organizations or problems. Each organization is unique and may require adaptation of the methodology to meet its specific needs. Furthermore, it is important to remember that successful implementation of Lean Six Sigma requires ongoing commitment to continuous improvement and organizational culture.

In conclusion, implementing Lean Six Sigma can be a valuable tool for improving efficiency and quality in business operations, and training and certification in the methodology can provide significant benefits for individuals' professional development and organizational culture. However, the effective implementation of Lean Six Sigma requires a significant commitment of time, resources, and leadership to overcome the challenges that may arise during the implementation process.

IMPLEMENTATION OF LEAN SIX SIGMA IN DIFFERENT INDUSTRIAL SECTORS: APPLICATIONS AND SUCCESS CASES

The implementation of Lean Six Sigma in the manufacturing industry has been extensively studied and documented. In this section, we will explore some success stories in different sub-sectors of the manufacturing industry.

Success Story in the Automotive Industry

Toyota is known for being one of the companies that pioneered the adoption of Lean philosophy in automobile production. The implementation of Lean Six Sigma at Toyota focused on continuous improvement of production processes to reduce waste and enhance the quality of vehicles.

Toyota's emphasis on continuous improvement has led the company to become one of the most efficient in automobile production. The company's focus on quality and efficiency has been so effective that it has become a model for other companies seeking to improve their processes.

Success Story in the Food Industry

The implementation of Lean Six Sigma in the food industry has focused on improving product quality, reducing production times, and minimizing waste. One successful example of Lean Six Sigma implementation in the food industry is the case of Heinz.

Heinz implemented the Lean Six Sigma methodology in its production plant in Scotland, which resulted in a 30% reduction in production costs and a 10% increase in production. The company also managed to improve the quality of its products, resulting in increased customer satisfaction.

Success Story in the Pharmaceutical Industry

The implementation of Lean Six Sigma in the pharmaceutical industry has focused on improving product quality and reducing production times. One successful example of Lean Six Sigma implementation in the pharmaceutical industry is the case of GlaxoSmithKline (GSK).

GSK implemented Lean Six Sigma in its production plant in the United Kingdom, which led to a 25% reduction in production cycle time and a 15% reduction in production costs. Additionally, the company managed to improve the quality of its products, resulting in higher customer satisfaction.

Implementation of Lean Six Sigma in the Service Industry

The implementation of Lean Six Sigma in the service industry has focused on improving efficiency and customer service quality. In this section, we will explore some success stories in different sub-sectors of the service industry.

Success Story in the Banking Industry

The implementation of Lean Six Sigma in the banking industry has focused on improving customer service efficiency and reducing waiting times. One successful example of Lean Six Sigma implementation in the banking industry is the case of Bank of America.

Bank of America implemented Lean Six Sigma in its mortgage lending department, which resulted in a 30% reduction in loan processing time and a 50% reduction in processing costs. Additionally, the company managed to improve customer satisfaction by providing faster and more efficient service.

Success Story in the Healthcare Industry

The implementation of Lean Six Sigma in the healthcare industry has focused on improving patient service efficiency and reducing waiting times. One successful example of Lean Six Sigma implementation in the healthcare industry is the case of Virginia Mason Medical Center.

Virginia Mason Medical Center implemented Lean Six Sigma in its healthcare department, which resulted in a 50% reduction in waiting times and a 30% reduction in healthcare costs. The company also managed to improve patient satisfaction by providing faster and more efficient service.

Success Story in the Logistics Industry

The implementation of Lean Six Sigma in the logistics industry has focused on improving supply chain efficiency and reducing production costs. One successful example of Lean Six Sigma implementation in the logistics industry is the case of DHL.

DHL implemented Lean Six Sigma in its logistics department, which resulted in a 20% reduction in production costs and a 30% reduction in delivery times. The company also managed to improve the quality of its services, resulting in higher customer satisfaction.

Challenges and Considerations in Implementing Lean Six Sigma

While Lean Six Sigma has proven to be an effective approach for improving quality and efficiency in different industrial sectors, it also presents some significant challenges and considerations that need to be addressed during its implementation.

Challenges in Implementing Lean Six Sigma

One of the main challenges in implementing Lean Six Sigma is the cultural change required to adopt this methodology. Employees must be willing to change their processes and ways of working to achieve Lean Six Sigma improvement goals. Additionally, implementing Lean Six Sigma may require a significant investment in time and resources.

Another challenge in implementing Lean Six Sigma is the need for careful planning and management to ensure that improvement objectives and outcomes are effectively achieved.

Considerations in Implementing Lean Six Sigma

When implementing Lean Six Sigma, it is important to consider the specific context and needs of each industry and organization. The methodology must be tailored and customized to meet the unique needs of each company.

Furthermore, training and skill development are essential for the success of Lean Six Sigma implementation. Employees need to be trained in Lean Six Sigma methodology and tools to ensure they can effectively participate in the continuous improvement process.

Conclusion

The implementation of Lean Six Sigma has proven effective in improving quality and efficiency in various industrial sectors, including manufacturing, healthcare, banking, logistics, and many more. The success stories presented in this book demonstrate the positive impact Lean Six Sigma can have on cost reduction, increased efficiency, and improved customer satisfaction.

However, there are also significant challenges and considerations that must be taken into account when implementing Lean Six Sigma. It is essential to customize and adapt the methodology to meet the unique needs of each organization. Additionally, cultural change and investment of time and resources can be challenging but necessary to achieve desired improvement results.

In summary, implementing Lean Six Sigma can be a powerful tool for improving quality and efficiency in any industrial sector. By addressing the challenges and considerations and personalizing the methodology to meet the organization's needs, a positive impact can be achieved in customer satisfaction, cost reduction, and efficiency improvement.

ABOUT THE AUTHOR

Industrial and Systems Engineer

Master in Administration with Quality and Productivity

3 Certifications

2 Technical Studies

More than 20 training courses

Winner of the Singapore Cooperation Programme ITE

Instructor, engineer, content creator, and writer.
Discover the modern industry with the most controversial engineer.
engr's Workshop

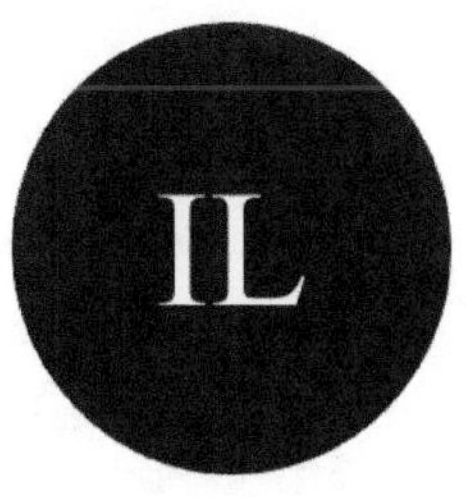

I. Laisequilla

Author / Engineer

RELATED BOOKS

The Bible of Industrial Engineering

From production management to process optimization, including methods and time engineering, this book covers the most important aspects of industrial engineering.

Available formats: physical, ebook, and audiobook

the all about Industrial Quality

FMEA, SPC, MSA, APQP, FMECA, Kaizen, Lean, ISO 9001, ISO 14001, ISO 45001, among others. Explained in an accessible and easy-to-understand manner.

Available formats: physical, ebook, and audiobook

the all about Industrial Methods

Lean Manufacturing, Six Sigma, Kaizen, TQM, Business Process Management. From classic methods to the most modern and emerging ones.

Available formats: physical, ebook, and audiobook